Form and Function of Insect Wings

Form and Function of Insect Wings

The Evolution of Biological Structures

Dmitry L. Grodnitsky

The Johns Hopkins University Press

Baltimore and London

© 1999 The Johns Hopkins University Press
All rights reserved. Published 1999
Printed in the United States of America on acid-free paper

9 8 7 6 5 4 3 2 1

The Johns Hopkins University Press
2715 North Charles Street
Baltimore, Maryland 21218-4363

www.press.jhu.edu

Figures 1.8, 2.4, and 3.6 first appeared in an article by the author
published in *Evolution* and are reprinted here with permission.
Figures 2.6, 2.7, 2.20–2.23, and 2.30 first appeared in articles by the
author published in the *Journal of Experimental Biology* and are
reprinted here with permission.

Library of Congress Cataloging-in-Publication Data will be found
at the end of this book.

A catalog record for this book is available from the British Library.

ISBN 0-8018-6003-2

To Natalia, My Mama

PRAISE FOR *WILD SPECTACLE*

"Here is Janisse Ray at her best—fully immersed in wilderness, immersed in friendship, immersed in parenthood. She engages with the world in a way that few can manage in this screened-off age. If there's a more open, honest, and appealing writer today, I've not met her."

— **Bill McKibben**, author of *Wandering Home: Long Walk across America's Most Hopeful Landscape*

"Janisse Ray's sense of wonder in the presence of the natural world permeates this collection of essays on how to love the Earth and measure the value of a life surrounded by the mother we all share."

— **Pam Houston**, author of *Deep Creek: Finding Hope in the High Country*

"Ray's richness of observation, clarity of expression, and moral purpose are in such balance that this book hums like a gyroscope in your hands. I read *Wild Spectacle* at the speed of Ray's travel across many landscapes. Read and reread it again to savor the scenes and sentences."

— **Melissa Fay Greene**, author of *No Biking in the House without a Helmet: 9 Kids, 3 Continents, 2 Parents, 1 Family*

"Wonderful. Janisse Ray has a heart the size of a manatee and the tenacity (and laugh) of a pileated woodpecker. She is incapable of not loving this world and all that is in it. If you don't yet know her work, today is your lucky day."

— **Rick Bass**, author of *For a Little While: New and Selected Stories*

"Seriously great. In its brilliantly detailed celebrations of geography, Janisse Ray's writing suggests Walt Whitman. Hers is a literary ambition that makes no pretenses to modesty."

— **Franklin Burroughs**, author of *Billy Watson's Croker Sack*

"Curious, humble, bright, and compelling. Whenever I read Janisse Ray, I come away feeling both moved and fortunate. She is one of America's best chroniclers of spiritual and physical wilderness. Her prose is as gorgeous as her mind is wise, and lands a necessary punch: how should a human enter a wild place?"

— **Megan Mayhew Bergman**, author of *Almost Famous Women*

"These seductive and diverse essays evoke wildness themselves, weaving narratives of community, love, and heroism. Ray writes with the heart of a poet and warrior, casting a spell that leaves us wanting to love and protect all that is wild. She urges us to remember what beauty there is in the world, and how much that world needs us."

— **Sheryl St. Germain**, author of *Fifty Miles*

"Janisse Ray's lyricism winds us to a heightened attentiveness to nature and calls with the clarity of dawn song throughout. I count her words as a wild blessing to the world."

— **J. Drew Lanham**, author of *Sparrow Envy: Field Guide to Birds and Lesser Beasts*

"Think about epiphany. Think about change. Think about the moments that make your face burn, your fingers tingle. *Wild Spectacle* is about those shocks, encounters that shift the way we see the world and ourselves in it. Although these essays range far and wide, Ray is the vortex around which everything spins. We journey with her into these places of salt and darkness and stone, into the sound and silence of conversations taken up and left off. These essays love wild people and trees, birds and wind, dirt and spiders. Reading this book, I remembered— if the water we drink is maybe older than the sun, then ancient magic pounds inside our skins, too."

— **Joni Tevis**, author of *The World Is on Fire: Scrap, Treasure, and Songs of Apocalypse*

"An urgent love letter to our wild places. Part poet, naturalist, and tour guide, Ray is a gifted observer. We finish this remarkable book brimming with gratitude and alive to the wild spectacles around us."

— **Beth Ann Fennelly**, author of *Heating and Cooling: 52 Micro-Memoirs*

"Janisse Ray doesn't explore nature so much as remind us of what we have forgotten. She is our Rachel Carson and our Walt Whitman—both fierce prophet and courageous teacher. She reminds us of what we have forgotten: that we are part of the beauty of the world, so long as we remember to be."

— **Mark Powell**, author of *Firebird*

"*Wild Spectacle* is a stirring book. To experience the truth of Thoreau's claim that wildness preserves the world, take these journeys with Janisse Ray. She is an exhilarating observer who explores untamed places where that shaping, animating energy is on vivid display."

— **Scott Russell Sanders**, author of *The Way of Imagination*

Contents

Preface and Acknowledgments   *xi*

1   Classification of Insects and Their Wingbeat Kinematics   1

   1.1.   Insect Taxonomy, *by Alexandr P. Rasnitsyn*   1
   1.2.   Classification of Insect Wing Kinematics   11

2   Wingbeats and Vorticity   17

   2.1.   General Points of Wing Movement and Deformation   17
   2.2.   General Aspects of Flapping-Flight Aerodynamics   21
   2.3.   Investigation of Airflow   27
   2.4.   Vorticity in the Flight of Morphologically and
       Functionally Two-Winged Insects   35
   2.5.   Vorticity at Posteromotorism   43
   2.6.   Vorticity in the Flight of Anteromotoric Functionally
       Four-Winged Insects   48
   2.7.   The Problem of Flow Separation from the
       Leading Edge   51
   2.8.   The Problem of Force Generation   58
   2.9.   Body Motion during Stroke Cycle   61
       2.9.1.   Motion of the Pterothorax   61
       2.9.2.   Motion of the Abdomen   64
   2.10. The Problem of Flight Power Regulation   67
   2.11. The Problem of Vortex Wake Geometry   71
       2.11.1. Near Vortex Wake   72
       2.11.2. Far Vortex Wake   73

2.11.3. Double Vortex Chains   78

2.12. Conclusions   79

3   Evolution of Flight in the Insect Orders   81

3.1. The Origin of Flight   82

3.2. Initial Wing Kinematics   86

3.3. The Origin of Major Kinematic Patterns   90

3.4. Grylloid Insects   94

3.5. Damselflies and Dragonflies   96

3.6. Hymenopterans and Rhynchote Insects   98

3.7. Beetles   101

3.8. Caddisflies, Moths, and Butterflies   104

3.9. Dipterous Flies   110

3.10. Functional Assessment of Some Elements of Wing Kinematics and Deformation   113

3.11. Conclusions   116

4   Problems of Endopterygote Insect Wing Functional Morphology   119

4.1. Topology of Veins and Furrows   121

4.2. Wing Morphology and Evolution in the Amphiesmenoptera   125

4.2.1. Nomenclature of Veins and Furrows   126

4.2.2. Transformation of Wing Planform   129

4.2.3. Transformation of Venation   132

4.2.4. Transformation of Furrow System   138

4.2.4.1. Caddisflies   141

4.2.4.2. Moths   142

4.3. Wing Morphology in Neuropterans, Scorpion Flies, and the Hypothetical Ancestor of the Amphiesmenoptera   143

4.4. Wing Morphology in Dipterous Flies   144

4.5. Wing Morphology in Beetles   148

4.6. Wing Morphology in Hymenopterans   151

4.7. Morphology and Functions of Furrows   153

4.8. Lepidopteran Wing Scale Covering   155

4.8.1. Size of Scales   156

Contents | ix

	4.8.2.	Covering Structure and Types	156
		4.8.2.1. Orientation of Scales	156
		4.8.2.2. Rows of Scales	157
		4.8.2.3. Covering Layers	158
		4.8.2.4. Differentiation of the Covering as a Function of Wing Zone	163
	4.8.3.	Functions of the Covering	166
4.9.	Conclusions		169

Appendix. Form and Function: A Review of General Concepts — 175

Basic Definitions — 176
Complementarity of Form-Generating Factors and Uncertainty of Explanations in Morphology: Recognition of the Problem — 180
Hierarchy of Faculties — 185
    The Principle of Evolutionary Stabilization of Functions — 186
    Another Formulation: The Principle of Minimum Change — 191
    Unavoidable Filling of Morphospace — 192
Multivariate Correspondence between Structures and Functions — 197
    Consequence 1: Adaptive Trade-off — 197
    Consequence 2: Faculty Clearance — 199
Optimality and Adequacy — 200
Conclusions — 204

References — *209*
Index — *253*

Preface and Acknowledgments

What is the essence of life—organisation or activity?
—Edward S. Russell, 1916

Functional morphology is one of the classic fields of biologic research that has been intensively developed throughout the last two centuries. Although it may have come of age, functional morphology has not become old. Throughout the twentieth century, it has been used to investigate phenomena predicted by evolutionary theory—the best-developed branch of theoretical biology. From the early 1900s until the present, the most popular research topic in functional morphology has been the directionality of evolution, the effect of natural selection. We now recognize that evolution is channeled by development, and that this process proceeds parallel to the constraints posed by natural selection. Hence, one of the crucial points of the theory of evolution is to determine the balance between ontogenetic and ecologic constraints. This is the main problem posed by this book.

The interaction of the two forces in biologic pattern formation is unlikely to be completely characterized in the framework of a single study. This is made even less likely because the material presented here approaches the problem from only one side, asking about the extent that the knowledge of function contributes to our understanding of the reasons for structural diversity. However, it is interesting to search for general biologic regularities from the viewpoint that theoretical considerations should be compared to results derived from experimental studies and natural observations. This is the current approach to the comparative functional morphology of insect wings.

Insect wings can be treated as flat structures. This feature makes the wings almost uniquely convenient objects for a survey of comparative morphology. The broad diversification of wing shapes and their constructions within insect class provide fruitful soil for analyses and reconstructions of historical trends.

The main function of insect wings is flapping flight. With respect to the distance covered during locomotion, energy expenses are fourfold lower in flying than in walking animals (Schmidt-Nielsen 1972). The high adaptive significance of this kind of locomotion has favored independent development of flight in several unrelated groups of animals. In insects, flight probably involves more diverse mechanisms than in any other animal taxon. Although disagreement remains on the way insects manage to stay aloft, certain regularities can clearly be described, thus explaining particular trends in insect evolution.

The first chapter presents a taxonomy and nomenclature of insects and provides a brief account of the main types of flapping flight in insects. The following three chapters are devoted to the functional morphology of wings. Usually, functional morphologists proceed from structure to function, first describing the diversity of construction and then showing how the structures work (see Pringle 1963; Brodsky 1994; Dudley 1999). Because the principal aim of this book is quite different, we will approach the object from the opposite direction, beginning with flight aeromechanics (chapter 2), next treating wing motion (chapter 3), and finally morphology (chapter 4). This approach is perceptually easier, because the diversity of the phenomena considered evidently increases from aerodynamics to wing construction. The book ends with an appendix containing a critical review of the ways in which functional morphology can contribute to general evolutionary theory, of the main principles of morphofunctional construction of organisms, and of the ways that these principles could define historical change of living beings.

I am not going to provide an overview of the structural diversity of the flight apparatus and subordinate organs in insects. Hence, the book lacks descriptions of the cuticular exoskeleton and thoracic muscles, and it does not contain analytic models for the energetics of flapping flight (for a comprehensive current reference, see Dudley [1999]). Rather, I have concentrated on disputable points in the functional morphology of wings.

Each chapter is intrinsically independent of the others and can be read separately. Potentially the book should be of interest to three groups of biologists. The first are the students of evolutionary process. Throughout the book and especially in the appendix, they will be exposed to concepts and data of purely Russian origin, both classic and modern. These concepts are mostly unknown in the West owing to the "splendid isolation" caused by obvious historical reasons. However, many of these concepts prove to be in good agreement with and complementary to current aspects of western biology. Hence, they may be of interest to western readers. Main evolutionary events accompanied by generation of higher taxa occur on large continents and not on islands. In a similar way, science develops more rapidly in such zones whenever it is provided with a rich diversity of concepts elaborated within different cultural traditions.

The second part of my audience consists of entomologists, who may be interested in the novel system of insect classification (the section written by A. P. Rasnitsyn). They will also encounter classification and simple explanation of the evolutionary origin of the major types of insect flight kinematics, as well as grounds for nomenclature and homology of insect wing furrows and venation. There is also a set of comparative data on functional morphology of the wings of endopterygote insects. The insect orders are distinguished mostly by the features of their mouthparts and flight systems; this makes the adaptive significance and corresponding bases of the evolutionary origins of the principal morphologic and kinematic traits of the orders especially important.

The third group of potential readers are experts in animal locomotion, both aerial and aquatic. They should be interested in the present description of the process of vortex generation associated with wing movements. They might also benefit from the general scheme of evolution of flight in the main orders of winged insects, beginning from a hypothetical origin of the wings and an initial state of their kinematics. Of course, the aeromechanics of flapping flight is far from common biological knowledge. Hence, I have tried to make explanations as easy to understand as possible and have supplied the text with supplementary illustrations.

I thank Peng Chai, Robert Dudley, Alexandr Emelianov, Carl Gans, Grigorij Kofman, Nikolaj Kokshaysky, Alexandr Rasnitsyn, Mikhail Sergeev, and Vladimir Svidersky for their critical comments on successive versions of the manuscript. Also helpful were communications and discussions with Georgij Liubarsky, Sergej Maslov, Viktor Novokshonov, Eric Pianka, Dmitry Shcherbakov, and Vladimir Zherikhin. Identified insect species used in comparative morphological study were offered by Anatolij Barkalov, Vladimir Blagoderov, Andrej Gourov, Vladimir Ivanov, Mikhail Kozlov, Dmitry Logunov, Sergej Lostchev, Viktor Marchenko, Agnia Mirzaeva, Olga Ovchinnikova, Liudmila Petrozhitskaja, Alexandr Rasnitsyn, and Boris Zakharov. Some living specimens for experiments were obtained from laboratory populations maintained by Olga Antonova, Irina Mikhailovskaja, and Alexandr Alekseev. Much technical assistance during all the stages of research was provided by Pahvel Morozov and Irina Mikhailova. The English in the manuscript was thoroughly checked and substantially revised by David Millard, who carried out a remarkable amount of work. Gratitude is also due to Charles Ellington, Michael Dickinson, Robin Wootton, and, especially, to Robert Dudley, who sent me literature sources now unavailable in Russia. Finally, I wish to express my debt to my wife, Irina, for her endless patience and understanding.

Form and Function of Insect Wings

1 | Classification of Insects and Their Wingbeat Kinematics

The organization and the nomenclature of higher insect taxa used in this book differ substantially from those generally accepted in current entomological literature. Therefore, before describing the diversity of insect flapping-flight kinematics, it is necessary to present a brief account of insect systematics.

1.1. Insect Taxonomy, *by Alexandr P. Rasnitsyn**

Insect systematics, especially at and above the taxonomic level of the order, has undergone continuous reorganization and improvement since Linné (1758). In particular, these changes have affected the content of the class Insecta (which initially corresponded more closely to the current concept of the phylum Arthropoda), the supraordinal taxonomic level, and the content of three of seven Linnean orders: Aptera, Neuroptera, and Coleoptera. The order Hemiptera was divested of thrips and in modern systems often has superordinal rank. The three other orders (Diptera, Lepidoptera, Hymenoptera) persist essentially intact since those remote times.

The revision of the class Insecta was time-consuming. Latreille (1810) separated crustaceans and arachnids from insects in the early nineteenth century and removed myriapods fifteen years later (Latreille 1825). However, it took more than a century to classify the entognathous orders Protura, Collembola, and Diplura as hexapod myriapods rather than true insects (Remington 1955;

* A. P. Rasnitsyn is the leader of the Arthropoda Laboratory, Paleontological Institute of the Russian Academy of Sciences, 123 Profsoyuznaya Str., Moscow, 117647 Russia; e-mail: rasna@glasnet.ru.

Handschin 1958; Sharov 1966), an inference that remains a subject of debate (cf. Rasnitsyn 1976a; Kristensen 1981b, 1992).

Clearing insect orders of extraneous elements required less time. Considerable progress was achieved by Brauer (1885), who left only two unnatural orders: an Orthoptera that included embiids and a Corrodentia that contained psocids, termites, and lice. Brauer correctly segregated the holo- and heterometabolan orders and, moreover, assembled the future palaeopteran, polyneopteran (except earwigs), paraneopteran, and mecopteroid orders. However, no formal taxa were proposed for these groups, and some of his orders (Orthoptera, Rhynchota, Neuroptera, etc.) are now considered to be superorders.

The next important step was made by Handlirsch (1906–8, 1911, 1925), who established numerous extinct orders, split the rest of the composite extant orders, and diversified the system of intercalar taxa between the subclass Pterygota and its component orders. Many of his new taxa still exist.

The first (and only) system of insects that won worldwide appreciation and incomparable longevity was developed by the Russian entomologist Martynov (1923, 1924a, 1925, 1938). This system, with or without minor modifications, is still in wide (though not universal) use. Martynov proposed that a basic dichotomy of the winged insects (Pterygota) segregates them into two divisions of the highest rank, depending on whether they can fold their wings at rest over their abdomen (Neoptera, comprising the majority of insect orders) or whether they cannot (Palaeoptera, namely, mayflies, dragonflies, and several Paleozoic orders close to Palaeodictyoptera). This idea is incorporated in various contemporary systems, either essentially intact (e.g., Ross 1965; Gillott 1980; Arnett 1993) or with a group discarded because of inferred paraphyly (Palaeoptera in Boudreaux [1979] and Kristensen [1981b, 1992]; Polyneoptera in Kukalova-Peck [1992]; Poly- and Paraneoptera are merged back into Exopterygota by Carpenter [1992]).

Despite the evident advantages of Martynov's system, his intellectual descendants realized a different and more radical trend in higher-level insect taxonomy. This group included students of Rohdendorf, who had studied under Martynov and who inherited and expanded Martynov's laboratory after his death in 1938. Specialized study of the taxonomic diversity of fossil insects, in close conjunction with the taxonomy and evolution of the living forms, prepared them to search for the basal insect dichotomies outside of the traditional approach.

The first step was made by Sharov (1966), who hypothesized that ancestral pterygotes (his new order Protoptera) were able to fold their wings at rest, albeit imperfectly. Further incremental development of Sharov's ideas produced an essentially new insect system and phylogenetic scheme (Rasnitsyn 1976a, 1980; Rohdendorf and Rasnitsyn 1980). The main feature of this approach is appreciation of the neopterous nature of ancestral pterygotes. Palaeoptera represent a disparate, polyphyletic group (Rasnitsyn 1997), and the basic pterygote dichotomy segregates Gryllones (Polyneoptera in Martynov's system) and Scarabaeones (the remainder of Pterygota).

Before presenting an overview of the most recent version of this system, another feature of the present taxonomic proposal should be explained. The difficulties encountered in working with an uncodified nomenclature inspired Rohdendorf's (1977; Rasnitsyn 1982) plan to codify the higher taxa (see also Rasnitsyn 1996), an approach that has been used in the proposed system. Although the following system does not require use of the codified nomenclature, I have little doubt that eventually this fully codified nomenclature will be appreciated. It provides the only way to stabilize the names of the higher taxa in a manner equivalent to that which has already been established with the names of the lower taxonomic ranks.

Class Scarabaeoda Laicharting, 1781 (= Insecta Linné, 1758). Insects.

Subclass Lepismatona Latreille, 1804 (= Thysanura Latreille, 1796, sensu lato). Wingless insects.

Order Machilida Grassi, 1888 (= Archaeognatha Börner, 1904; = Microcoriphia Verhoeff, 1904; + Monura Sharov, 1957). Bristletails. Possibly the sister group to other insects (= Ectognatha Stummer, 1891). Upper or Middle Carboniferous, unless Devonian (Labandeira et al. 1988; Shear et al. 1984), to present; now worldwide (fig. 1.1).

Order Lepismatida Latreille, 1804 (= Thysanura Latreille, 1796; sensu stricto = Zygentoma Börner, 1904). Silverfish. Middle Carboniferous to present; now worldwide.

Subclass Scarabaeona Laicharting, 1781 (= Pterygota Lang, 1888). Winged insects.

Infraclass Scarabaeones Laicharting, 1781.

Cohors Paoliiformes Handlirsch, 1906.

Order Paoliida Handlirsch, 1906 (= Protoptera Sharov, 1966). Middle and Late Carboniferous (mostly Namurian); North America, Europe, rarely in Siberia. Supposedly ancestral to all other winged insects, known only from wings (fig. 1.2).

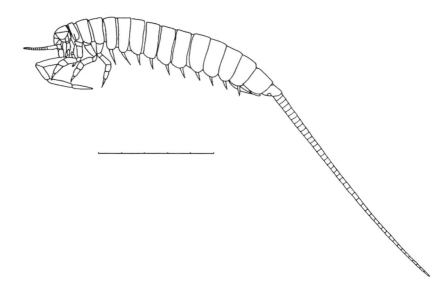

Figure 1.1. *Dasyleptus brongniarti* Sharov (Dasyleptidae), a representative of the Machilida, a proposed sister group of other insects. Upper Permian of southern Siberia. Scale 5 mm. (After Rohdendorf and Rasnitsyn 1980.)

Cohors Libelluliformes Laicharting, 1781 (= Subulicornia Latreille, 1806; = Hydropalaeoptera Rohdendorf, 1968).

Superorder Ephemeridea Latreille, 1810.

Order Triplosobida Handlirsch, 1906 (= Protephemerida Handlirsch, 1906). Late Carboniferous of Europe. Poorly known group.

Order Syntonopterida Handlirsch, 1911. Lower Middle Carboniferous to Lower Permian of Europe and North America. Includes Syntonopteridae and Bojophlebiidae. Archaic ancestors of mayflies (fig. 1.3).

Order Ephemerida Latreille, 1810 (= Ephemeroptera Haeckel, 1896). Mayflies. Lower Permian to present; worldwide.

Superorder Libellulidea Laicharting, 1781 (= Odonata Fabricius, 1792, s.l.; + Geropteroidea Brodsky, 1994). Dragonflies.

Order Eugeropterida Riek & Kukalova-Peck, 1984 (= Geroptera Brodsky, 1994). Carboniferous of Argentina. Very archaic, known only from wings.

Order Meganeurida Handlirsch, 1907. Lowermost Middle Carboniferous to Lower Cretaceous; worldwide, but prefer warm climates. Probably ancestral to Libellulida (fig. 1.4).

Order Libellulida Laicharting, 1781 (= Odonata Fabricius, 1792; sensu stricto

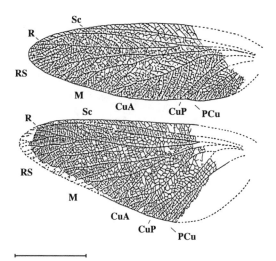

Figure 1.2. Zdenekia grandis Kukalova (Paoliidae), a representative of the proposed ancestors of winged insects, known only by fossil wings. Lowermost middle carboniferous of Czechia. For designations of veins see section 4.2.1 and figure 4.5. Scale 20 mm. (After Kukalova 1958.)

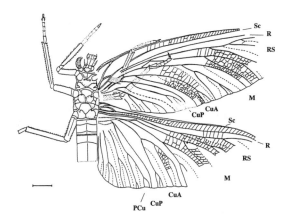

Figure 1.3. Bojophlebia prokopi Kukalova-Peck (Bojophlebiidae), a representative of the ancestors of mayflies. Middle carboniferous of Czechia. Designations as in figure 1.2. Scale 20 mm. (After Kukalova-Peck 1985.)

= Neodonata Martynov, 1938). True dragonflies and damselflies. Late Triassic to present; worldwide.

Cohors Cimiciformes Laicharting, 1781. Former Paraneoptera and related archaic orders.

Superorder Caloneuridea Handlirsch, 1906. Suggested ancestral group to both Paraneoptera (via Hypoperlida) and Oligoneoptera.

Order Cnemidolestida Handlirsch, 1906 (= Cnemidolestoidea Handlirsch, 1906; sensu Handlirsch, 1937; = Gerarida Handlirsch, 1906; sensu Rohden-

Figure 1.4. Arctotypus sp. (Meganeuridae), a representative of the proposed ancestors of damselflies and dragonflies. Upper Permian of northern Russia. Scale 40 mm. (After Rohdendorf and Rasnitsyn 1980.)

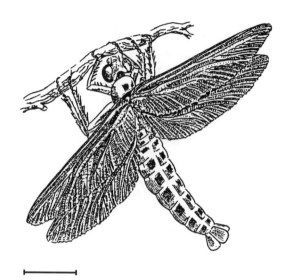

dorf, 1977). Middle to Late Carboniferous; Europe and North American (equatorial belt of the Carboniferous).

Order Blattinopseida Bolton, 1925. Middle Carboniferous to Late Permian; Eurasia and North America.

Order Caloneurida Handlirsch, 1906 (= Caloneurodea Martynov, 1938). Middle Carboniferous to Late Permian; Eurasia and North America.

Order Zorotypida Silvestri, 1913 (= Zoraptera Silvestri, 1913). Miocene to present; worldwide but patchy in warm climates, including oceanic islands. Supposedly a sister group of holometabolous insects (Rasnitsyn, forthcoming).

Superorder Hypoperlidea Martynov, 1928.

Order Hypoperlida Martynov, 1928. Uppermost Lower Carboniferous to Upper Permian of Eurasia and North America. A suggested ancestral group to both Paraneoptera and Palaeodictyopteroidea (fig. 1.5).

Superorder Dictyoneuridea Handlirsch, 1906 (= Palaeodictyopteroidea Rohdendorf, 1961).

Order Cacurgida Handlirsch, 1906 (former Paoliida without Paoliidae and Evenkidae). Lower Middle to Upper Carboniferous of Europe and North America. Supposedly ancestral to the rest of the superorder.

Order Dictyoneurida Handlirsch, 1906 (= Palaeodictyoptera Goldenberg, 1854). Lower Middle Carboniferous to Late Permian, with highest abundance during Middle and especially Late Carboniferous; widespread in

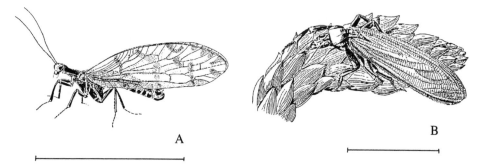

Figure 1.5. A, Tshekardobia osmylina Rasnitsyn (Ampelipteridae); *B, Synomaloptila longipennis* Martynov (Synomaloptilidae), suggested ancestors of rhynchote and paleodictyopteroid insects. Lower Permian of Urals. Scale 20 mm. (After Rohdendorf and Rasnitsyn 1980.)

Northern Hemisphere, but common only in Eurameria, most probably owing to its warmer climate; absent on southern continents. Ancestral to the following orders:

Order Mischopterida Handlirsch, 1906 (= Megasecoptera Brongniart, 1893; + Archodonata Martynov, 1932). Middle and Upper Carboniferous, Lower and very rare in Upper Permian; worldwide although rare, except in Eurameria and westernmost Angarida.

Order Diaphanopterida Handlirsch, 1906 (= Diaphanopterodea Handlirsch, 1906). Lowermost Middle Carboniferous to Lower Permian; North Eurasia and North America.

Superorder Psocidea Leach, 1815 (= Psocopteroidea Rohdendorf, 1961).

Order Psocida Leach, 1815 (= Psocoptera Shipley, 1904; = Copeognatha Enderlein, 1903). Booklice. Lower Permian to present; worldwide with particular diversity in tropics. Ancestral to lice and birdlice, as well as to thrips.

Order Nyrmida Leach, 1815 (= Mallophaga Nitzsch, 1818). Birdlice. Worldwide; no fossil record.

Order Pediculida Leach, 1815 (= Anoplura Leach, 1815; = Siphunculata Latreille, 1825; = Phthiriaptera Haeckel, 1896). Lice. Worldwide; recorded since Eocene.

Superorder Thripidea Fallén, 1814.

Order Thripida Fallén, 1814 (= Thysanoptera Haliday, 1836). Thrips, including Lophioneuridae. Lower Permian to present; worldwide.

Superorder Cimicidea Laicharting, 1781 (= Rhynchota Burmeister, 1835).

Figure 1.6. Palaeomantina pentamera Rasnitsyn (Palaeomantiscidae), a representative of the supposed ancestors of endopterygote insects. Lower Permian of Urals. Scale 10 mm. (After Rasnitsyn 1980.)

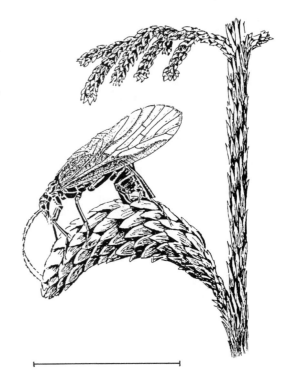

Order Cimicida Laicharting, 1781 (= Hemiptera Linné, 1758; = Homoptera Leach, 1815; + Heteroptera Latreille, 1810). Lower Permian to present; worldwide.

Cohors Scarabaeiformes Laicharting, 1781 (= Holometabola Burmeister 1835; = Endopterygota Sharp, 1899; = Oligoneoptera Martynov, 1925).

Superorder Palaeomanteidea Handlirsch, 1906.

Order Palaeomanteida Handlirsch, 1906 (= Miomoptera Martynov, 1927). Upper Carboniferous to Upper Jurassic; North Eurasia and North America, rarely also in Australia and South Africa. Supposedly ancestral group to all other holometabolans (fig. 1.6).

Superorder Scarabaeidea Laicharting, 1781 (= Coleopteroidea Handlirsch, 1903).

Order Scarabaeida Laicharting, 1781 (= Coleoptera Linné, 1758). Beetles. Lower Permian to present; worldwide. The largest (richest in species number) order of living beings.

Order Stylopida Stephens, 1829 (= Strepsiptera Kirby, 1813). Eocene until present; worldwide. Relationship to beetles is questionable (cf. Kristensen 1992).

Superorder Myrmeleontoidea Latreille, 1802 (= Neuropteroidea Handlirsch, 1903).

Order Raphidiida Latreille, 1810 (= Raphidioptera Navas, 1918). Snakeflies. Lower Jurassic to present; after Lower Cretaceous not found at southern continents.

Order Corydalida Leach, 1815 (= Megaloptera Latreille, 1802). Alderflies and dobsonflies. Upper Permian to present; worldwide. Fossils only in Eurasia, except for a single South African Triassic finding.

Order Myrmeleontida Latreille, 1802 (= Neuroptera Linné, 1758). Lacewings and ant lions. Lower Permian to present; worldwide.

Order Jurinida M. Zalessky, 1928 (= Glosselytrodea Martynov, 1938). Lower Permian to Middle Jurassic; worldwide. Jurassic fossils found only in Asia. This order and Palaeomanteida are the only extinct holometabolous orders.

Superorder Papilionidea Laicharting, 1781 (= Mecopteroidea Martynov, 1938).

Order Panorpida Latreille, 1802 (= Mecaptera Packard, 1886; = Mecoptera Comstock & Comstock, 1895; + Paratrichoptera Tillyard, 1919a; + Paramecoptera Tillyard, 1919b). Scorpion flies. Lower Permian to present; worldwide, but with most living groups extratropical.

Order Phryganeida Latreille, 1810 (= Trichoptera Kirby, 1813). Caddisflies. Lower Permian to present; worldwide but more abundant in temperate climate.

Order Papilionida Laicharting, 1781 (= Lepidoptera Linné, 1758). Butterflies and moths. Lower Jurassic to present; worldwide.

Order Muscida Laicharting, 1781 (= Diptera Linné, 1758). Flies. Middle Triassic to present; worldwide.

Order Pulicida Billberg, 1820 (= Aphaniptera Kirby, 1826; = Siphonaptera Latreille, 1825). Fleas. Eocene to present; true fleas are worldwide. Questionable fossils from the Upper Jurassic and Lower Cretaceous of East Asia and Australia indicate the sister-group relationship of fleas to the other Papilionidea (Rasnitsyn 1992b).

Superorder Vespidea Laicharting, 1781 (= Hymenopteroidea Handlirsch, 1903).

Order Vespida Laicharting, 1781 (= Hymenoptera Linné, 1758). Wasps, bees, and ants. Middle or Upper Triassic to present; worldwide. Supposedly a sister group of other holometabolans excluding Palaeomanteida.

Infraclass Gryllones Laicharting, 1781 (Polyneoptera Martynov, 1923).

Superorder Blattidea Latreille, 1810 (= Dictyoptera Clairville, 1798).

Order Eoblattida Handlirsch, 1906. Middle and Upper Carboniferous of Britain, France, and the United States; probably confined to the warmest climate. Supposedly ancestral to other blattidean orders.

Order Blattida Latreille, 1810 (= Blattodea Brunner, 1882). Roaches. Lower Middle Carboniferous to present; worldwide but most abundant in warm climates. Possibly ancestral to both termites and praying mantids.

Order Termitida Latreille, 1802 (= Isoptera Brullé, 1832). Termites. Early Cretaceous to present; warm temperate, subtropic, and especially tropic zones (generally within 40° north and south) of all continents.

Order Manteida Latreille, 1802 (= Mantodea Burmeister, 1835). Praying mantids. Lower Cretaceous to present; living mantids are termophilous. For the oldest fossils, see Gratshev and Zherikhin (1993).

Superorder Perlidea Latreille, 1802 (= Plecopteroidea Martynov, 1938).

Order Grylloblattida Walker, 1914 (= Notoptera Crampton, 1915; = Grylloblattodea Brues & Melander, 1932; + Protorthoptera Handlirsch, 1906; = Paraplecoptera Martynov, 1925; + Protoperlaria Tillyard, 1928). Middle Carboniferous to present; worldwide in Paleozoic, now amphi-Pacific relicts in Holarctic. Ancestral to other perlidean orders (fig. 1.7).

Order Perlida Latreille, 1810 (= Plecoptera Burmeister, 1835). Stoneflies. Upper Lower Permian to present; worldwide although rare in warmer climates.

Order Forficulida Latreille, 1810 (= Dermaptera DeGeer, 1773; + Protelytroptera Tillyard, 1931). Earwigs. Lower Permian to present; worldwide.

Order Embiida Burmeister, 1835 (= Embioptera Shipley, 1904). Web spinners. Oligocene (or, possibly, Upper Cretaceous, if this is the age of the Burmese amber) to present; living embiids are worldwide in warmer climates.

Superorder Gryllidea Laicharting, 1781 (= Orthopteroidea Handlirsch, 1903).

Order Gryllida Laicharting, 1781 (= Orthoptera Olivier, 1789). Grasshoppers and crickets. Upper Carboniferous to present; worldwide.

Order Phasmatida Leach, 1815 (= Phasmatodea Burmeister, 1835). Stick insects. Late Permian to present; fossils comparatively rare, found in Eurasia and Australia only. Living forms common in all tropic regions, rare in warm temperate ones.

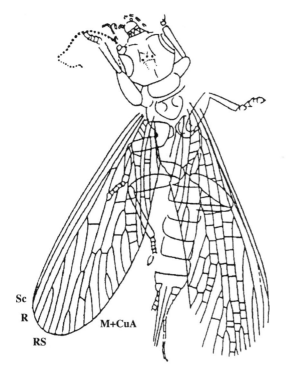

Sc
R
RS
M+CuA

Figure 1.7. *Blattogryllus karatavicus* Rasnitsyn (Blattogryllidae), the ancestor of stoneflies and related orders. Upper Jurassic of southern Kazakhstan. Designations as in figure 1.2. Scale 10 mm. (After Rasnitsyn 1976b.)

Order Mesotitanida Tillyard, 1925 (= Titanoptera Sharov, 1968). Middle and Late Triassic only; Central Asia, Australia, and eastern Europe.

1.2. Classification of Insect Wing Kinematics

The identification of insect orders relies considerably on features of the wings. Insects can have one or two wing pairs, with one pair larger than the other (either fore- or hindwings can be longer, depending on taxon). The wings can be hardened (elytra), membranous or leathery, fringed, scaled or hairy; veins and cells can be numerous or strongly reduced. From this, one is able to suggest that the wings—and flying, as their main function—played quite an important role in the evolution of insects. Indeed, after its origination, flight evolved in almost all insect orders. The historical change of flight imposed constraints on the general appearance of subordinate taxa, thus considerably channeling their evolution. Hence, the traditional classification and nomenclature of insect orders was based mainly on features of wing apparatus, especially since the wings are more frequently found as fossils than any other part of an insect's body (Martynov 1938).

Insects possess the most diverse wing structure and kinematics of all flying animals. Shvetz and coworkers (1979) classified insect flapping flight according to the character of flight trajectory and recognized three main flight types: straight, hovering, and flitting (the last a mixture of the first two). Ellington (1984) suggested a differentiation of flight types according to the position of the stroke plane, which can be horizontal, inclined, or vertical (see sec. 2.1). Nevertheless, a detailed classification (i.e., tree- or gridlike system) of flight modes was never developed, as most authors concentrated on intensive studies of model species and devoted little attention to the comparison of different groups. The early attempt of Rohdendorf (1949) dealt not with flight itself, but with the morphology of the wing apparatus. Brodsky (1988, 1994) listed flight modes, but did not bring them into mutual correspondence; that is, he only typologized the object. Moreover, his flight types coincided with insect taxa and therefore lacked independent significance. As will be shown further, partitioning of functional diversity does not concur with the taxonomic system: several flight modes can occur in one group of insects and can even be demonstrated in the different behavioral repertoires of a single insect species. For example, dragonflies can use up to four distinct kinematic patterns (sec. 3.5).

In the subsequent text, the term *wing couple* signifies a forewing and a hindwing taken together, whereas *wing pair* indicates a left and a right wing, that is, either forewings or hindwings. The newly suggested classification (fig. 1.8) is based mainly on the terminology of Shwanvitch (1946, 1948, 1949). The difference is that in the cited works the terms *anteromotoric* and *posteromotoric* have purely morphologic meaning, designating the greater size of either the fore- or the hindwings, respectively. As follows from the material presented subsequently, the relative size of the two wing pairs is determined by a functional character, which consists in a correlation between the beginning of the stroke cycle of the fore- and hindwings. The only known exception are some stoneflies, which start each stroke with their forewings, although their hindwings are enlarged.

The proposed scheme includes only those insects that are capable of active aerial flight. Not presented, for example, are some species of caddisflies (Phryganeida: Limnephilidae: Baicalini) (Kozhov 1963, 1973) and bugs (Cimicida: Helotrephidae, Pleidae) (Popov 1971) that have lost their hindwings together with the ability to fly. The term *morphological wing* merely corre-

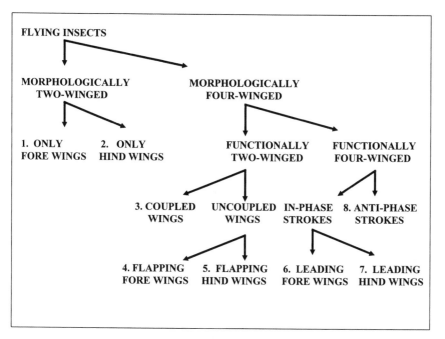

Figure 1.8. Classification of the main types of insect wing strokes.

sponds to a well-developed wing or elytron that is at least open during flight. The term *functional wing* designates a wing that performs strokes. Thus, cetoniine beetles (Scarabaeida: Scarabaeidae: Cetonia, Potosia) are regarded as two-winged, because they never open their elytra. Small wing rudiments or strongly modified wing derivatives (as in Nemopteridae) unable to evoke any vorticity are not considered. The main intergroup differences reflect the number of working wings and interrelationships between the fore- and hind-wing pairs.

Morphologically two-winged insects have a single pair of well-developed wings. Only the forewing pair is present in dipterous flies (Muscida), some Syntomidae (Papilionida), tiny wasps (Vespida: Mymarommatidae, Mymaridae), Nemopteridae (Myrmeleontida), scale insects (Cimicida: Coccoidea), and some mayflies (Ephemerida). Remarkably, these insects are often of small size—smaller than the mean size characteristic of their groups in general. For example, morphologically two-winged mayflies are the smallest among all the ephemeropterans. The same is true of the previously mentioned wasps and scale insects. The order Diptera, in general, also includes rather small insects.

Many species possess only hindwings. These include some beetles (Scarabaeida: Staphylinidae, Cetoniinae), twisted-wing parasites (Stylopida), some walking sticks (Phasmatida), mole crickets (Gryllida: Gryllotalpidae), pygmy grasshoppers (Gryllida: Tetrigidae), pygmy mole crickets (Gryllida: Tridactylidae), and earwigs (Forficulida).

Among the insects that have four true wings, functionally two- and four-winged forms are recognized. Most functionally two-winged species possess fore- and hindwings connected as two couples. During flight each couple operates as a single wing. This condition is typical of many taxa, including the Cimicida, and most of the Phryganeida, Papilionida, and Vespida.

Several functionally two-winged species are characterized by four unpaired wings that all remain open during flight. Nevertheless, flapping is performed by only one of the two wing pairs. Thus, some diurnal butterflies (Papilionida: Papilionidae: *Parides neophilus, Papilio* spp., *Ornithoptera* spp.) can fly with immovable hindwings, so that all the aerodynamic force necessary for flight is generated by the forewings, which work at frequency values that are unusually high for butterflies (>25 strokes per second). Remarkably, these butterflies are also able to fly in the usual lepidopteran manner, with paired wings (Grodnitsky et al. 1994). Operating the wings at high-frequency values is energetically more expensive (Pennycuick 1975). At the same time, neither the thorax construction, the musculature, nor the wing articulations of these butterflies are adapted to work at frequencies higher than 5–15 Hz. This functional modification, clearly maladaptive for flight, can be explained in the following way. The functionally two-winged papilionids are unpalatable, with hindwings bearing brightly colored warning spots. The butterflies are most vulnerable to predators when they are gathering nectar and flitting from flower to flower. To avoid attacks, these butterflies keep their hindwings fully open and stationary, so that the bright pattern is clearly seen. Thus, the butterflies use extra energy but enhance their chances for survival (Grodnitsky et al. 1994).

Tiger beetles (Scarabaeida: Carabidae: Cicindelinae) provide an example of functionally two-winged insects that fly with the help of their hindwings. In flight their elytra are open, fixed, and supported from below by the forelegs and middle legs (Schneider 1975; Grodnitsky and Morozov 1994). The same kinematics also can be observed rarely in maneuvering dragonflies (Rüppell 1989).

The fore- and hindwing pairs of functionally four-winged insects flap relatively independently of each other. In the majority of species, the fore- and hindwings display in-phase kinematics, producing a phase shift that is either constant or changing in accordance to a definite law, and is short in comparison with the entire stroke cycle. Normally the stroke amplitude is considerably larger in the hindwings than in the forewings.

Functionally four-winged anteromotoric (with leading forewings) in-phase kinematics is characteristic of scorpionflies (Panorpida), lacewings (Myrmeleontida), primitive moths (Papilionida: Micropterigidae, Eriocraniidae), primitive caddisflies (Phryganeida: Rhyacophilidae), and stoneflies (Perlida) (Ellington 1984; Grodnitsky and Kozlov 1985; Ivanov 1985a; Brackenbury 1992, 1994; Grodnitsky and Morozov 1992, 1994). At the beginning of the stroke, the hindwings of these insects follow the forewings, then outrun them approximately in the middle of the downstroke, and, finally, returning in the upstroke, meet the forewing pair at the highest point of the trajectory to start a new working cycle.

Note, however, that the inclusion of stoneflies in this group is somewhat problematic. According to Brodsky (1982, 1985, 1986a), the stoneflies *Isogenus nubecula* Newm. and *Allonarcys sacchalina* Klap. are anteromotoric, whereas *Taeniopteryx nebulosa* L. and *Isoptena serricornis* Pict. are posteromotoric. This may explain why Brodsky (1988, 1994) designated stonefly strokes as quasi-synchronous. Nevertheless, the other two Perlida species are evidently anteromotoric (see fig. 3.2*B*) (Grodnitsky and Morozov 1994). This kinematic type is presumably also specific to webspinners (Embiida), snakeflies (Raphidiida), and alderflies (Corydalida). Anteromotoric strokes also may occur rarely in dragonflies (Rüppell 1989; Sato and Azuma 1997).

In-phase flappers that begin strokes with their hindwings are called *posteromotoric*. They include dragonflies (Libellulida), most grasshoppers, crickets and mantids (Gryllida), and, in all likelihood, cockroaches (Blattida). While flying, their hindwings are permanently leading (Pringle 1957; Neville 1960; Zarnack 1983; Alexander 1984; Kammer 1985; Schwenne 1990; Brackenbury 1990, 1991a, 1992; Robertson and Reye 1992; Robertson and Johnson 1993; Wortmann and Zarnack 1993).

The wingbeat kinematics of beetles is somewhat different; nevertheless, it should be classified as in-phase functionally four-winged with a leading hind pair. Although all the wings begin the stroke cycle almost simultaneously, in the majority of the beetles, the stroke amplitude of the elytra is strongly reduced and does not exceed 30–60° (Schneider 1975; Grodnitsky and Morozov 1994).

A peculiar type of functionally four-winged kinematics has been observed in anisopterous dragonflies; their fore- and hindwings perform antiphase strokes (phase shift equals approximately a half-cycle) (Chadwick 1940; Neville 1960; Norberg 1975; Alexander 1984; Rüppell 1989). Contrary to the previous pattern, the hindwings begin to move from the lowest point of their trajectory, when the forewings start from the top, and vice versa. This flight mode is related to particular behavioral responses: the same insects are capable of in-phase flight as well.

The selection factors that channeled the evolution of the diversity of insect flapping flight are unclear and have not yet been widely discussed. A causal explanation for the origination of the main kinematic modes is based on the results of comparative studies of the vorticity generated by insect wings during flight. Data on vorticity created during the stroke cycle are presented in chapter 2, and an evolutionary scenario of flight change is discussed in chapter 3.

2 | Wingbeats and Vorticity

The principal function of wings is flight; therefore, a functional assessment of the morphology and evolution of insect wings initially requires insight into the mechanisms of aerial locomotion. This chapter examines the major phenomena that occur in the air near a flying insect, and correlates these phenomena with the principles of force generation that have been suggested for insect flapping flight. Before describing insect flight dynamics, it is necessary to give a brief account of how the wings move, and to explain the conventional ways in which force sufficient for flight is created.

2.1. General Points of Wing Movement and Deformation

In an insect-bound coordinate grid, the wing tip trajectory reveals a closed loop of various curvatures and shapes that may change according to flight situation. The upstroke trajectory is normally behind the downstroke path. As a first approximation, the wings can be considered to move in a plane containing the wing hinge. This plane is called the *stroke* plane. It is perpendicular to the longitudinal body axis in many species, and is inclined to the body axis at an angle of 30–60° in others (fig. 2.1*A*, *B*). While hovering the stroke plane is often close to horizontal (Ellington 1984).

The stroke cycle begins at the start of downward wing movement from the highest point of the wing trajectory. According to relevant literature (Pringle 1957; Weis-Fogh 1973; Ellington 1984; Kammer 1985; Betts 1986a; Wootton 1990) and original data, the wingbeat cycle is quite similar in very different insects. The main feature of the wing stroke during the forward flight of all volant animals is its asymmetry (Lighthill 1969; Spedding 1992), which is necessary to generate the aerodynamic force to stay aloft. This force is cre-

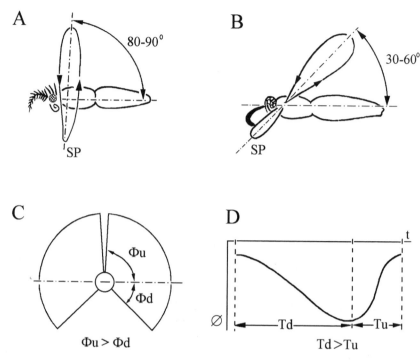

Figure 2.1. Common features of wingbeat kinematics: (*A, B*) shape of trajectory and stroke plane types: O-shaped vertical (stroke plane angle close to 90°) (*A*) and 8-shaped inclined (stroke plane angle is acute) (*B*); (*C*) relation between upper (Φu) and lower (Φd) parts of stroke cycle; (*D*) relation between downstroke (Td) and

ated owing to a difference in airspeeds over and under the wing. The force is positive if the airspeed is greater above the wing than below it (during downstroke), and negative when the opposite occurs (normally during up-stroke). The asymmetry of strokes in flapping flight is required to make the value of the positive lift larger than that of the negative, so that the difference between downstroke and upstroke impact is positive and sufficient for flight (Alexander 1968; Cloupeau et al. 1979; Nachtigall 1985; Brodsky 1988; Sunada et al. 1993a).

The stroke cycle is asymmetric in several respects. First, the downstroke is normally incomplete, because the wings do not touch in the lower part of their trajectory. Thus, the lower portion of the stroke is much shorter than the upper part (fig. 2.1C). Second, the downstroke usually lasts longer than the upstroke (fig. 2.1D), so the downstroke-to-upstroke duration ration, although variable, cannot be less than 1. Third, the aerodynamic angle of attack (the

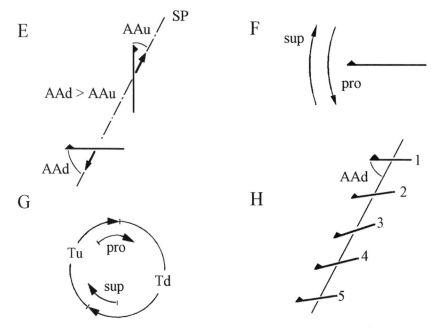

upstroke (Tu) durations; (E) difference between mean angle of attack during down-stroke (AAd) and upstroke (AAu) (direction of wing movement is shown by arrows); (F) pronation (pro) and supination (sup); (G) correspondence between durations of downstroke (Td), upstroke (Tu), pronation (pro), and supination (sup); (H) angle of attack change characteristic of the "swinging-edge mechanism" (1–5 consequent frames of the stroke cycle beginning). ∅ = wing positional angle, SP = stroke plane.

angle between the mean wing chord and the airflow) is always larger during downstroke in comparison with the second half-stroke (fig. 2.1E). To adjust the angle of attack, the wings rotate about their longitudinal (i.e., hinge-tip) axis. The rotation of the leading edge downward takes place near the upper-most point of the trajectory and is called *pronation*. Upward leading-edge rotation is called *supination* and is observed near the lowest point of the stroke cycle (fig. 2.1F). Typically pronation begins before and stops after the wings have passed the highest point. The same is true of supination in relation to the lowest point (fig. 2.1G).

An inclined stroke plane appears as a secondary feature in many unre-lated insect orders. It strengthens the stroke asymmetry, because in straight flight the wing airspeed and corresponding generated force are significantly increased during the downstroke and diminished in the upstroke (Woot-ton 1990).

In both extreme trajectory points during wing rotation, a peculiar movement termed *swinging-edge mechanism* can be observed (Nachtigall 1979, 1985; Zanker 1990a; Grodnitsky and Morozov 1994). Instead of changing the angle of attack monotonically, the wing first performs a redundant rotation through a turn considerably greater than necessary to achieve the angle of attack required for the forthcoming downstroke. Next, the wing turns in the opposite direction and achieves the angle of attack that is kept approximately constant during the further half-stroke (fig. 2.1*H*).

The wings of all flying insects display similar features of deformation during the stroke cycle (Wootton 1981). During most of the cycle, the wings are slightly cambered, being almost flat (fig. 2.2*A*). In species with a secondarily enlarged ano-jugal zone, the hindwings assume the shape of a gutter directed backward or backward and aside. In broad-winged species (mantids, grasshoppers, and butterflies) (Brackenbury 1990, 1991a, 1992; Wootton 1995), a slightly convex wing profile can be observed. Insects with wings of usual width only occasionally demonstrate a convex profile (Grodnitsky and Morozov 1994).

Because of their construction, insect wings are much more deformable during supination than pronation (Ennos 1988; Wootton 1993). In supination and upstroke, the wings become strongly concave. This achieves lower angle of attack values and thus diminishes the negative lift that accompanies the upward movement of the wings. In all insects, the concave upstroke profile is formed, in part, by smooth wing bending. Additionally, in a majority of species, smooth bending is amplified by more or less sharp inflections of the wing in strictly definite zones, termed *plicas* by Martynov (1924b), *sutures* by Emelianov (1977), *flexion* and *fold lines* by Wootton (1979), *furrows* by Mason (1986), and *hinges* by Banerjee (1988). I shall subsequently refer to them as "furrows." (See chapter 4 for a discussion of their nomenclature, arrangement, and functions.) At this point, it is sufficient to state that there are generally two lines of deformation on both the fore- and the hindwing. In generalized insects, one of the deformation lines is stretched along the remigial zone of the wing, near the median vein. In many specialized forms, the vein is lost, whereas the deformation line is retained. Another line lies on the border between the remigium and the clavus. Also, in insects with coupled fore- and hindwings, the zone of wing joint may occasionally serve as a concave deformation line (fig. 2.2).

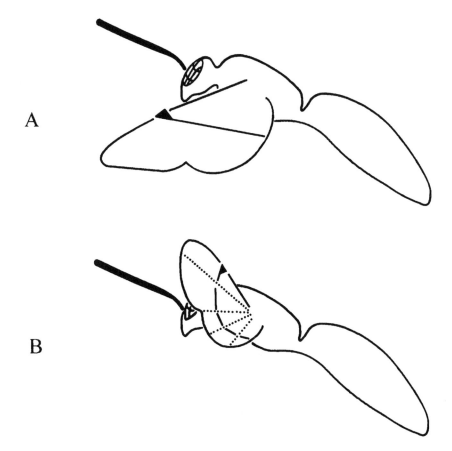

Figure 2.2. Common features of wing section shape during downstroke (A) and upstroke (B). Concavities are shown with dotted lines.

2.2 General Aspects of Flapping-Flight Aerodynamics

Locomotion in fluids is strictly limited by physical requirements and is thus definitely less diverse than terrestrial locomotion. The main principles of force generation have much in common in all flying and swimming animals and were considered in several general reviews of animal aerial and aquatic locomotion (Kokshaysky 1974; Ellington 1984, 1995; Rayner 1986; Spedding 1992; Dudley 1992, 1999; Dickinson 1996). Advances in the field of biomechanics of natural locomotion in fluids can be attributed in large part to the well-developed theory of flying and swimming of man-made vehicles. Although this theory is unable to explain all the phenomena observed in nature,

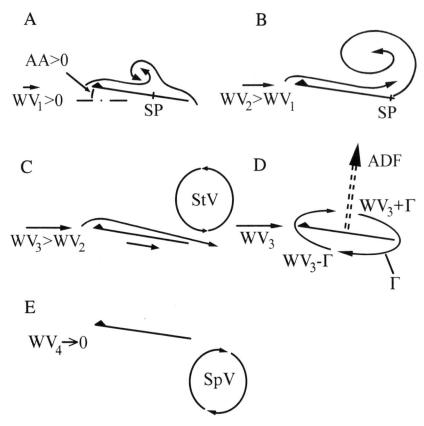

Figure 2.3. Successive stages of vortex formation near aircraft wings during initial start (*A*), movement on runway (*B*), takeoff (*C*), flight (*D*), and landing (*E*). AA, angle of attack; WV, velocity of contrary airflow; SP, separation point; StV, starting vortex; Γ, circulation; ADF, aerodynamic force; SpV, stopping vortex. Triangles mark the upper surface of the wing leading edge.

it clearly facilitates research, suggesting various mechanisms for useful force generation.

Let us consider aircraft aerodynamics. Because of a positive angle of attack, the first moments of movement are accompanied by air acceleration near the leading and trailing edges of the wing and generation of rarefaction on its upper side. This results in two flows, one coming from the leading, the other from the trailing edge. As the flows meet, they separate from the wing surface and curl up, forming a vortex (fig. 2.3*A*). Next, the point of separation of the flow initially located between the leading and trailing edges shifts backward with the growth of magnitude and rotation velocity of the vortex above the wing (fig. 2.3*B*). This process culminates when the point of separation coincides with the trailing edge, and a starting vortex is generated (fig. 2.3*C*). From this moment on, air moves about the wing without separation, faster above the wing and slower below it.

The growth period of the starting vortex is accompanied by an increase in aerodynamic force applied to the wing, its vertical component called *lift*. A measure of the difference of airspeeds above and below the wing is called *circulation*. It can be represented in an abstract way, as if a "bound" vortex rotates around the wing, accelerating air over and decelerating it under the wing (fig. 2.3*D*). Ideally the respective momenta of the starting and bound vortices are equal. Circulation continues around the wing during flight and departs from the wing at landing, at which time a stopping vortex is formed that rotates in the direction opposite to the starting one (fig. 2.3*E*). The starting and stopping vortices form the vortex wake of flying aircraft.

In animals, the force necessary to fly is generated as wings accelerate the surrounding air. As the wings do this, they are said to be "aerodynamically active" (or efficient). The resulting air movement constitutes a vortex wake analogous to that of an aircraft. Because wake shapes reveal peculiarities of mechanisms of force generation, studies of vorticity are given priority in the aeromechanics of aerial locomotion, and have in fact defined the main success achieved by this science during the last decade (Hummel and Goslow 1991; Rayner 1991; Ellington 1995). Data have been published on the flow around flying birds (Kokshaysky 1979, 1982; Kokshaysky and Petrovsky 1979a, 1979b; Spedding et al. 1984; Spedding 1986, 1987), bats (Rayner and Aldridge 1985; Rayner et al. 1986), and insects (Chance 1975; Martin and Carpenter 1977a, 1977b; Ellington 1980; Brodsky and Ivanov V. D. 1983a, 1984; Brodsky 1985, 1986b, 1988, 1990, 1991; Brodsky and Grodnitsky 1985; Ivanov 1990; Grodnitsky and Morozov 1992, 1993, 1995; Ellington 1995; Ellington et al. 1996; Grodnitsky and Dudley 1996; van den Berg and Ellington 1997a, 1997b; Willmott et al. 1997).

The following terms are used to describe the three-dimensional structure of vortex wakes. A vortex tube consists of a cylinder-shaped space volume in which a rotatory air movement takes place. Ideally the air rotates in the direction tangential to the cylinder surface. A vortex tube cannot end in air and must be closed either to a solid surface (e.g., a wing) or to itself. In the latter case, a self-closed loop called a *vortex ring* is generated (fig. 2.4). The ring has two axes of symmetry: central (fig. 2.4*A*) and circular (fig. 2.4*B*). A planar section of the vortex tube reveals a vortex (fig. 2.4*B*). Thus, the vortex is a

A
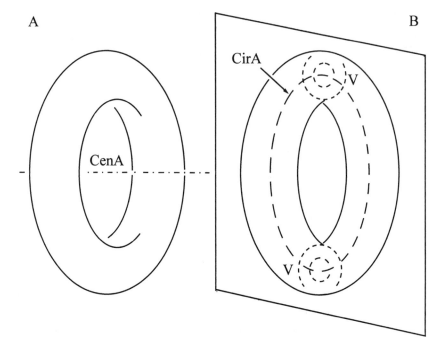
B

Figure 2.4. Vortex ring (*A*) and its section (*B*) by plane, containing central symmetry axis (CenA) of the ring. V, two-dimensional vortex; CirA, circular axis of symmetry.

two-dimensional structure, whereas the tube and ring are three-dimensional. Since flow rate and pressure in liquid and gas are inversely proportional, vorticity represents a zone of low pressure.

The three-dimensional structure of the vortex wake reflects the main difference between animal flapping flight and the flight of man-made aircraft (Kokshaysky 1974). In an airplane, the starting vortex constitutes a cross section of the starting vortex tube (also called the *starting bubble*) generated close to the trailing edge and stretched from one tip of the wing to the other (fig. 2.5*A*). After the aircraft has passed a distance equal to several wing chords (Saffman 1992), the bubble separates from the trailing edge and stays behind the wing, so that only a pair of vortex tubes remain attached to the tips (fig. 2.5*B*) (Prandtl and Tietjens 1957). Cross sections of these tubes are called *tip vortices.* While decelerating, the wing generates a stopping vortex bubble (fig. 2.5*C*). As a result, an aircraft produces a single, infinitely stretched vortex ring from the beginning to the completion of its flight. The aerodynamic force necessary to fly is created entirely from the bound vortex.

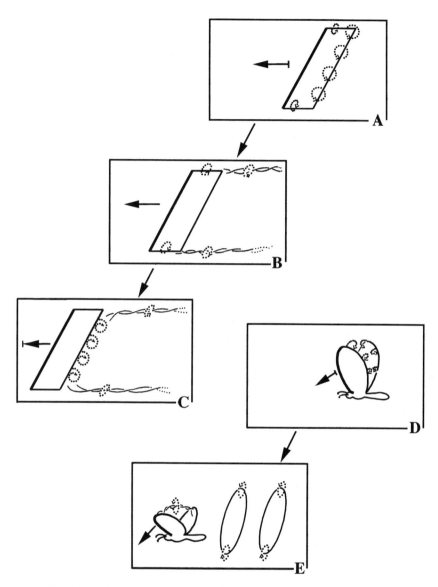

Figure 2.5. Three-dimensional structure of vortex wake behind aircraft plane (A–C) and flapping insect (D, E): (A) start of movement; (B) flight; and (C) landing. (D) Beginning, and (E) continuation of the stroke with general scheme of the wake generated during previous stroke cycles. Arrows show the direction of wing movement; dotted lines indicate rotational flow (vortices). Leading edges of the wings are thickened.

At the beginning of an insect stroke cycle, the flapping wing (or each wing couple, in functionally two-winged species) also generates a starting vortex tube (fig. 2.5*D*). After a short time (before the wings' positional angle reaches 60°), these two tubes merge and create a single U-shaped bubble, analogous to the starting bubble of an aircraft (fig. 2.5*E*). Ellington (1995) termed this bubble the *combined vortex*.

At the bottom of their trajectory, the wings decelerate and swing around. This movement produces a stopping vortex tube that together with the U-shaped bubble, forms a self-closed vortex ring. Because the upstroke is also aerodynamically active, tip vortices come from the wings during this stroke period. These tip vortices merge into the ring generated during downstroke, so that no new independent vortices are formed in the upstroke. The following stroke cycle generates another vortex ring. Thus, the entire aerodynamic wake of a flying insect consists of a series of mutually parallel rings (Ellington 1980, 1984; Grodnitsky and Morozov 1992, 1993, 1995; Grodnitsky and Dudley 1996), independent of a kinematic pattern of the wings. This is the simplest scheme of in-stroke vortex dynamics. As will be shown further, different aerodynamic phenomena can take place in insect flight, and the resulting shape of the wake can be quite varied. Nevertheless, the most important idea to be understood at this point is that, in respect to the generated vorticity, the downstroke in insects corresponds to the entire flight period in aircraft.

This correspondence is the most consistent characteristic of natural flapping flight. In aircraft, the starting vortex remains far behind the wing after the initial stage of movement. Thus, it does not affect the forces generated by the aircraft wings. On the other hand, in insects the starting vortex is produced many times within each second of flight. At least at the initial stroke phase, the starting vortex is located close to the body and wings. Owing to this proximity, it can influence the process of force generation. As theory predicts and experiments demonstrate, the presence of a vortex bubble near a wing results in additional vertical force (Polhamus 1971; Savchenko 1971; Belotserkovsky et al. 1974a, 1974b; Kokshaysky 1974; Savage et al. 1979; Edwards and Cheng 1982; Wu et al. 1992; Dickinson and Götz 1993). Thus, performance of the wing can be improved by means of its interaction with the region of the vortex wake closest to it; these interactions can be quite diverse (see Brodsky 1994; Ellington 1995; Ellington et al. 1996; Willmott et al. 1997).

Uldrick (1968) suggested that flying and swimming animals possess adaptations that enable them to control energy dispersal to the vortex wake and extract energy back from the wake, adjusting the wings (or fins) to the gener-

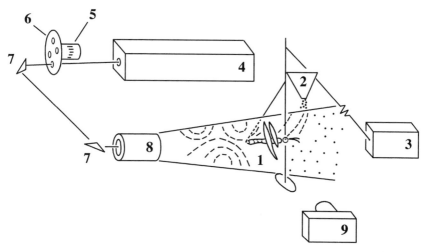

Figure 2.6. The scheme of installation for study of airflow around tethered flying insects: 1, specimen; 2, container with dust; 3, vibrator; 4, laser; 5, motor; 6, rotating disc with openings; 7, prism; 8, cylindrical lens and condensor; 9, camera.

ated vortices. Developing Uldrick's suggestion further, I argue that the main kinematic patterns of insect wings described in chapter 1 originated as adaptations to diminish the dissipation of energy during the formation of starting vortex bubbles.

2.3. Investigation of Airflow

The generation and dynamics of vortices can be most clearly revealed through the method of flow visualization, which consists in adding particles to the air that follow its movement and render it visible. Various substances that have been used in this procedure include smoke (Brodsky and Ivanov V. P. 1974, 1975; Brodsky 1991; Ellington 1995; Ellington et al. 1996; van den Berg and Ellington 1997a, 1997b; Willmott et al. 1997), fine solid particles, termed *dust* by researchers (Kokshaysky 1979; Ellington 1980; Brodsky and Ivanov V. D. 1983a, 1984; Ivanov 1990; Dickinson and Götz 1996), and small helium-filled soap bubbles (Spedding et al. 1984; Rayner and Aldridge 1985; Rayner et al. 1986; Spedding 1986, 1987).

Grodnitsky and Morozov (1992, 1993, 1995) used spores of lycopodium (club moss) as the visualizing material in their experiments. The spores were poured on a flying insect from a vibrating container (fig. 2.6). With this system, the experimenter can regulate the amount of dust in the camera viewing field by adjusting the amplitude of vibration.

A laser was used as the source of illumination. A cylindrical lens modified the light beam into a flat stripe approximately 1 mm wide. Illumination with a flat beam is a convenient element of the visualization procedure in studies of flow around insects and artificial models of their wings (Vogel 1965; Vogel and Feder 1966; Maxworthy 1979; Brodsky and Ivanov V. D. 1983a, 1984; Spedding and Maxworthy 1986; Dickinson and Götz 1996). Pictures obtained with this method represent sections made by the light plane in the three-dimensional vortex flow, because the images of only those dust particles located in the beam plane at the moment of frame exposure are fixed on film. By changing the position of the beam relative to the insect, one can obtain pictures of the flow in various parts of the surrounding space during different phases of the stroke cycle. Sixteen different light plane positions were used in five aspects: lateral, frontal, front-aside, and top-aside (fig. 2.7). Afterward, the three-dimensional flow structure was reconstructed from the two-dimensional representations, and its stroke cycle dynamics was described. In this technique, the photographs do not record the central part of the vortices, because dust is thrown away from the area of rotational air motion. The method described permits the study of flow only about insects with a relatively low wingbeat frequency, because high frequencies increase the speed of the ambient air, with a corresponding thinning of dust particle images below the limits of film resolution.

The resolution capacity of the method can be increased by illuminating the field of vision with a glimmering laser beam (Grodnitsky and Morozov 1993). The frequency of glimmering can be changed with a simple mechanical stroboscope, consisting of an electric motor and a perforated metal disc (fig. 2.6). Adjusting the rotational speed of the circle enables one to arbitrarily modulate the beam, that is, to change the duration of light impulses, as well as to vary the impulse/pause duration ratio. An impulse frequency is used that is equivalent to the wingbeat frequency of the insects under study. The resulting visualized flow picture becomes static and can be photographed at prolonged ($1/30$–$1/8$ s) exposure time, providing considerably more informative images. A small decrease in impulse frequency makes the flow picture move, so that a slow succession of vortices can be observed at low magnification.

Grodnitsky and Morozov (1992, 1993, 1995) obtained data on visualization from experiments conducted in still air. It is critical to know how the absence of any opposing wind affects the motion of wings. A comparative

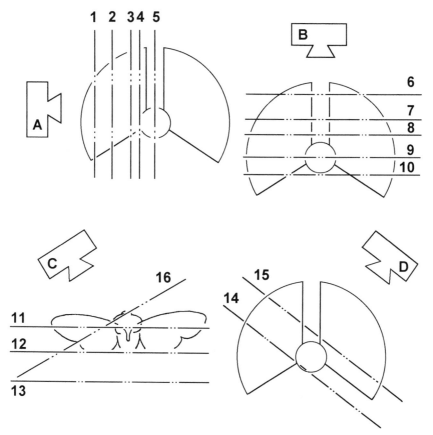

Figure 2.7. Different variants of allocation of light plane (1–16) and the aspects of filming: (*A*) from aside; (*B*) from above; (*C*) from front and above; (*D*) from aside and above.

study of the tethered flight of two lepidopteran species in flowing and in still air demonstrated that in airflow, wings are brought 10–20° closer to the plane containing the wing hinges, and perpendicular to the longitudinal body axis (fig. 2.8). The difference is quite subtle; therefore, studies of flight in still air can be considered adequate at first approximation. Ivanov (1985a, 1990) derived identical conclusions from his investigations of the natural and experimental flight kinematics of caddisflies.

An evolutionary interpretation of the data on geometry and stroke dynamics of vortices, as well as on wingbeat trajectory and wing shape change, is only possible with sufficient comparative material. Broad data for comparison can be obtained only by examining insects in tethered flight, but the

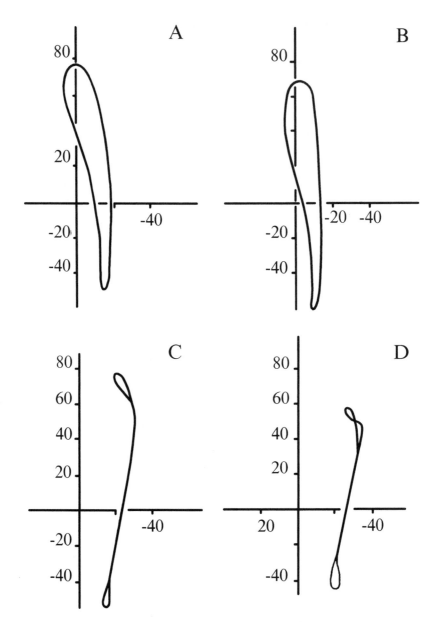

Figure 2.8. Trajectories of forewing tip of the codling moth *Laspeyresia pomonella* (A, B) and European skipper *Thymelicus lineola* Ochs. (Hesperiidae) (C, D) during tethered flight in airflow (A, C) and in still air (B, D). Coordinates: horizontal, angle between wing and the plane perpendicular to the longitudinal body axis; vertical, angle between wing and horizontal.

use of tethered insects in the study of natural flight needs additional support. A tethered specimen cannot be used to evaluate the correlation between its weight and generated aerodynamic force; the information perceived by its visual and wind-sensitive organs is far from natural, and its center of gravity, which normally undergoes periodical vertical oscillations, is immovable. Therefore, it is, of course, preferable to study free flight (Norberg 1975; David 1978; Baker and Cooter 1979a, 1979b; Ellington 1984; Azuma et al. 1985; Alexander D. E. 1986; Betts 1986c; Azuma and Watanabe 1988; Betts and Wootton 1988; Ennos 1989a; Rüppell 1989; Brackenbury 1990, 1991a, 1991b, 1992, 1994; Dudley 1990; Dudley and Ellington 1990a; Sunada et al. 1993b; Dudley and Srygley 1994; Wakeling and Ellington 1997a, 1997b, 1997c).

However, there are aspects of flight that cannot presently be investigated using free-flying insects. These include, for example, operation of musculature, nervous ganglia and nerves, as well as the vorticity around body and wings. Only a few studies of this type have been published, being carried out on single species (Kutsch et al. 1993; Grodnitsky and Dudley 1996). However, these studies did not contribute considerably to the existing understanding of flapping flight. Therefore, tethered flight will remain one of the primary technical approaches in this field. It is necessary, recognizing the constraints imposed by tethering, to try to define the relationship of the experimental results to natural phenomena. Wingbeat frequency in tethered flight is generally less than that of insects in free flight (Baker et al. 1981; Kutsch and Stevenson 1981; Feller and Nachtigall 1989; Grodnitsky 1992a), and even at similar wingbeat frequencies, tethered specimens generate less thrust (Gewecke 1983). Thrust production is probably reduced because the wings in tethered specimens perform almost harmonic strokes (Td/Tu close to 1) (Grodnitsky 1992a), whereas the downstroke-to-upstroke ratio in free-flying insects always exceeds 1 (Ellington 1984; Betts and Wootton 1988).

Owing to the difference of kinematic parameters, tethered flying insects often do not counterbalance their own weight (Gewecke 1975, 1983; Baker et al. 1981; Ward and Baker 1982; Götz 1987). Consequently, energy expenditure during tethered flight is less intensive (Casey and May 1983; Casey et al. 1985; Feller and Nachtigall 1989), temperature characteristics are decreased (Heinrich 1993), and tethered insects are able to fly for longer time periods (Cockbain 1961). In addition, experimental and free insects, even those belonging to the same species, can behave in different ways during turns (Yager and May 1990), and the motion of wings after attachment acquires unnatural asymmetry (Möhl 1988). Wing kinematics is also dependent on the experi-

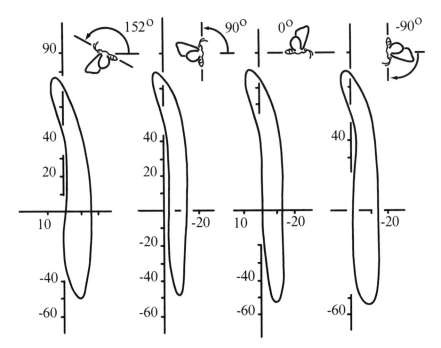

Figure 2.9. Trajectories of forewing tip of the codling moth L. *pomonella* at different orientation of the body. Coordinates as in figure 2.8. (From Grodnitsky and Kozlov 1987.)

mental procedure and different results can be obtained under different conditions (Hollick 1940; Heisenberg and Wolf 1979; Ellington 1984; Zanker 1990b).

It is essential that different insects react to various experimental conditions in different ways. For instance, in locusts a so-called lift-control reaction has been described (Gettrup and Wilson 1964; Gettrup 1965), in which the tethered flying specimen keeps the generated lift constant irrespective of the angle between its body and horizontal (body angle), at the expense of changes in wingbeat kinematics. In several other studied species, by contrast, arbitrary variation of the body angle does not affect the angle between the longitudinal body axis and stroke plane, nor the general shape of the wing tip path (fig. 2.9). Any change of the flying specimen's body angle results in a corresponding change of its weight (force with which it affects the tether). The curve describing the dependence of vertical force on the angle of inclination of stroke plane to horizontal (stroke plane angle) resembles a sinusoid (fig. 2.10). This indicates a lack of the lift-control reaction and shows that rotation in the sagittal plane does not influence the angle between the stroke

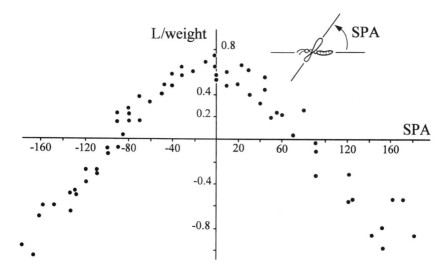

Figure 2.10. Dependence of weight-specific lift (L/weight) on the stroke plane angle (SPA, °) in tethered noctuid moth *Heliothis armigera* Hbn., flying in still air. (From Grodnitsky 1992b.)

plane and the vector of aerodynamic force generated by the wings, or other stroke parameters (amplitude, angle of attack, etc.) that determine the force magnitude.

The shape of the sinusoid is dependent on the opposing wind (fig. 2.11): its descending half shrinks along the horizontal, resulting from an increase of air speed about the body and wings, which produces a corresponding increase in aerodynamic force. Maximum lift generated by tethered insects can exceed the weight of the individuals. When lift values approximated the weight of the experimental specimens, their body angles were close to those observed in the same species during natural flight (30–50°). Therefore, while it can be concluded that experimental conditions disturb wingbeat kinematics, it can also be inferred that this disturbance is not strong, because the generated force is comparable with the force necessary for free flight.

On the other hand, some data indicate an absence of significant differences between free and fixed flight, concerning flight speed (Niehaus 1981), temperature regime of the pterothorax (Jungmann et al. 1989), the value of generated force (Hollick 1940), and, in particular, aerodynamically significant kinematic parameters.

Available data on the deformation of wings during free flight (Dalton 1975;

Figure 2.11. Dependence
of weight-specific lift
(L/weight) on stroke plane
angle (SPA, °) at different
velocities of contrary wind
(WV, m/s): (A) pentato-
mid bug *Antheminia aliena*
Reut.; (B) pine sawyer
Monochamus sutor L.;
(C) yellow jacket wasp
Dolichovespula silvestris
Scop. (From Grodnitsky
1992b.)

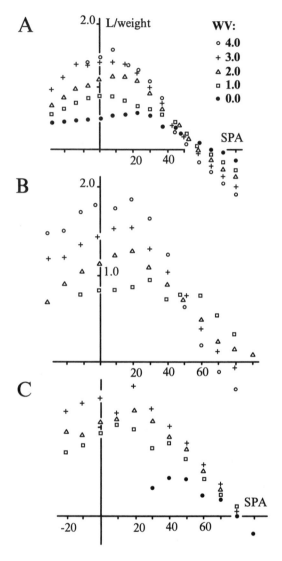

Mareš and Lapàček 1980; Betts 1986c; Brackenbury 1990, 1991a, 1991b, 1992, 1994; Wootton 1990; Brackenbury J. H. 1994) does not differ fundamentally from results obtained in experiments with tethered insects (Grodnitsky and Morozov 1994): in both situations the same qualitative patterns of wing pro-file change take place.

Comparison of vortices generated around insects in free and in tethered flight (Grodnitsky and Morozov 1993; Grodnitsky and Dudley 1996) also shows that the data generally coincide, because vortex size and stroke motion relative to the body are similar. It should be mentioned, however, that the natural phenomenon is more diverse than its experimental approximation. Thus, in free specimens single strokes of low amplitude lead to the appearance of a separate vortex wake after each wing pair. In tethered flight, double wakes never occur, since the wingbeat amplitude is often even higher than in free-flying specimens (Grodnitsky 1992a).

The final impression is that the differences between free and tethered flight of insects are not connected with flight as a physical phenomenon, but rather reflect the behavioral reactions of specimens to the unnatural conditions of the experiment. Consequently, the least cooperative tethered insects are those that have complex behavioral repertoires: dragonflies, bees and wasps, and butterflies (except skippers). Accordingly, kinematic modifications can cause shifts in the relationships between particular phases of the stroke cycle and in corresponding changes in quantitative correlation between aerodynamic processes, which take place during these phases. The qualitative essence of the phenomenon, however, remains basically unchanged. In any case, the researcher has an opportunity to choose among the entire diversity of insect behavioral reactions, and to experiment with species in which tethered flight approximates the natural phenomenon. Overall, it can be concluded that presently there are no data demonstrating the impossibility of the investigation of qualitative aspects of insect flight on tethered specimens, and that studies of tethered flight provide a reasonable approximation of natural aerial locomotion.

From the considerations presented in the foregoing, one should not conclude that studies of insect flapping flight must be restricted purely to tethered flight research. Instead a combination of diverse methods of investigation is desirable within the limits defined by the nature of the particular flight aspect under study.

2.4. Vorticity in the Flight of Morphologically and Functionally Two-Winged Insects

For any newly investigated problem, disagreements are more natural than consensus. The description of the aerodynamic processes that take place during active flight of insects is no exception. This is especially true because the three-dimensional wake structure is rarely observed directly, but instead is reconstructed, and therefore it should be treated as hypothesis rather than fact. The latter circumstance causes numerous controversies in the debate on

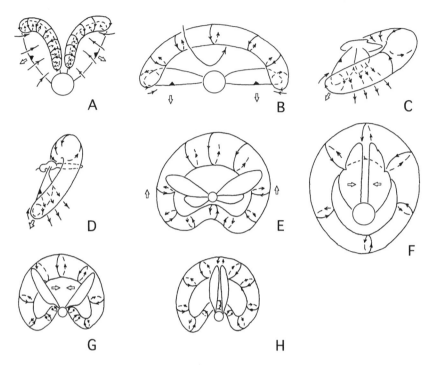

Figure 2.12. Formation of vortex tubes during stroke cycle of functionally and morphologically two-winged insects. (*A*) Beginning and (*B*) middle of downstroke (frontal views). (*C*) End of downstroke in lepidopterans and rhynchote insects; (*D*) the same in crane fly (lateral view). (*E*) Middle and (*F*) end of upstroke in lepidopterans, rhynchote insects, and muscid flies (frontal views). (*G*) Middle and (*H*) end of upstroke in crane fly (frontal views).

vortex wake spatial structure. I shall consider available information, beginning from the most simple case, on morphologically and functionally two-winged insects: dipterous flies, lepidopterans, hymenopterans, and rhynchote insects.

The stroke cycle begins with the motion of wings downward from the highest point of their trajectory. When the trailing edges begin to move away from each other, starting vortex tubes are formed over them. In bugs and dipterous flies, flow separation and the corresponding generation of starting vortices is seen along the entire length of the trailing edges (fig. 2.12*A*). Two bubbles are formed, closed to the forewing tips by their distal ends. The proximal ends of the bubbles are closed to the trailing edges at the beginning of departure of the wings. The diameter of the transverse section and the rotation speed of the tubes are not uniform: vortices near the body are weaker, whereas the most intensive vortex generation occurs near the rapidly moving distal parts of the wings.

As the wing speed increases, the starting tubes grow in size and rotate more rapidly. They elongate simultaneously, and their proximal ends shift along the trailing edges toward the insect body. Consequently, the tubes approach each other and merge, forming a single U-shaped vortex bubble over the body of the flying insect. The bubble is closed by its ends to the tips of the forewings (fig. 2.12B, C); in all likelihood, this occurs when the wings achieve a positional angle between 45 and 60°. After the bubble has been formed, the distribution of internal vorticity becomes more uniform. As a result, longitudinal sagittal sections of the flow reveal a large dorsal vortex (fig. 2.13). The U-shaped bubble accelerates air over the body of the insect, directing it backward.

The described events of formation of starting tubes (fig. 2.12A) has not been found in butterflies and moths. Unlike those of other studied insects, lepidopteran wings tightly touch each other over the body at the beginning of the stroke cycle. This seems to be connected with the specificity of lepidopteran kinematics in the beginning of the stroke (see chap. 3). Motion of the wings in these insects is accompanied by the gradual departure of the trailing edges, which are separated in a manner analogous to a zipper, so that the point of connection of the right and left wings glides along the trailing edge toward the insect abdomen. This motion was termed *peel* by Ellington (1984). Peel permits the starting tubes generated by the left and right wings (or wing couples) to merge early in the stroke cycle. In butterflies and moths, the formation of a U-shaped vortex bubble presumably occurs at the very beginning of the cycle. A similar pattern is observed in caddisflies, which have wing morphology and starting stroke kinematics similar to moths (Ivanov 1990).

In lepidopterans and bugs, the flow created by the wings brings the U-bubble backward as the first half of the working cycle continues. Consequently, at the end of the half-stroke the dorsal vortex is no longer over the body, but is located instead above the tip of the abdomen (fig. 2.12C). At an equivalent time in dipterous flies, the dorsal vortex retains its initial position approximately over the wing hinges (fig. 2.12D). Hence, it can be concluded that the U-bubble is not brought backward in these insects.

In all investigated species, supination at the end of the first and at the beginning of the second half-stroke results in clearly visible airflow thrown off the dorsal wing surface (fig. 2.14). The observed pattern resembles the flow

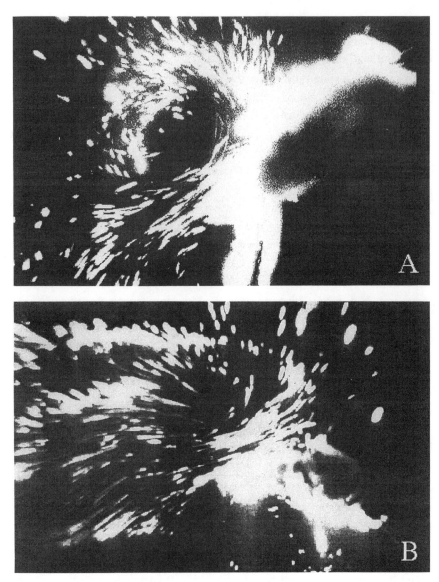

Figure 2.13. Vortex over the body during downstroke: (*A*) pentatomid bug, (*B*) muscid fly. Lateral view, light plane position 5 (see fig. 2.7).

Figure 2.14. Fanlike air throw-off during supination of the wings (insect flying from left to right). Lateral view, light plane position 2 (see fig. 2.7).

around a fan during the reversal of its motion, which explains why this phenomenon was designated *fanlike throw-off* (Grodnitsky and Morozov 1993). This effect has been seen in all the insect species studied.

The supinational turn of the wings begins before and finishes after the second half-stroke begins, because the downward motion of the trailing edge proceeds after the wing tips start moving upward. Since the aerodynamic processes that determine the flow character of the entire wing take place near the trailing edges, the flow picture typical of the second half-stroke is observed not just after the beginning of the upstroke, but somewhat later, after the end of supination. In fact, downstroke appears to be longer for trailing edges than for wing tips. Therefore, the borderline between the first and second half-strokes should be confined to the trailing edges, not to the wing tips' movement reversal. As a result, separation of the stopping vortex tubes from the trailing edges is delayed and arises after the beginning of the upstroke, coinciding in time with the end of the fanlike throw-off.

The stopping tubes are located beneath the wings. Their proximal ends close to the trailing edges and their distal ends to the wing tips. Consequently,

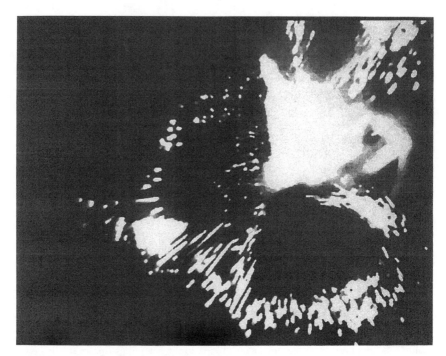

Figure 2.15. Vortex under insect during upstroke (insect flying from left to right). Lateral view, light plane position 3 (see fig. 2.7).

the starting U-bubble unites with the stopping tubes via the wing tips, forming a vortex ring, which is disrupted under the abdomen. By analogy with the downstroke, the stopping tubes accumulate vorticity that comes from the wings, grow in a proximal direction, and eventually meet under the insect abdomen, forming a self-closed vortex tube. This process occurs in the majority of studied species (fig. 2.12E), but the ventral tubes in different species unite at varying times.

In bugs, butterflies, moths, and muscid flies, the vortex ring closes soon after the completion of supination, before the wings achieve the positional angle of 90°. Hence, during the remaining part of the upstroke, a ventral vortex is seen under the abdomen of these insects. The ventral vortex has a smaller diameter than the dorsal vortex, and rotates in the opposite direction (fig. 2.15). Thus, the wings operate nearest to the vortex ring during the second half of the stroke cycle. The upper two-thirds of the ring are presented by the starting tube; the bottom third is the stopping tube. A new separate ring, corresponding to the upstroke, does not arise. Otherwise, the flow pictures taken from above would contain not one, but two sequential, tandem-located vortices on each side of an insect body. The wings pronate before the stroke cycle is fully completed, causing a fanlike air throw-off from the wings similar to that which occurs during supination (fig. 2.14). Air is thrown off the lower wing surface during pronation, whereas during supination it is thrown off the upper side.

Bocharova-Messner and Aksiuk (1981) described a peculiar tunnel formed by the ano-jugal zones of the hindwings of the butterfly *Gonepteryx rhamni* at the end of each stroke cycle. As they approach each other, the wings push air backward through the tunnel, and the insect is propelled forward by a jet impulse. Brodsky (1994) and Willmott et al. (1997) supported this hypothesis. However, the results of dust visualization show that the air is thrown away from the lower surfaces of the wings, rather than from the space between them. This can be seen only in dorsal views with a horizontal light plane, not in the lateral views used in the previously cited articles. In addition, the formation of a tunnel is impossible, unless the left and right wings clap at the upper part of their trajectory, which is a relatively rare type of kinematics in insects (see chap. 3). By contrast, fanlike throw-off does not require that the wings meet over the body. The flow that originates from the throw-off brings the completely formed, self-closed vortex ring (fig. 2.12F) backward, so that the ring loses its connection with the wings. The backward travel of the ring provides room for the formation of new starting tubes at the beginning of the following stroke.

The aerodynamic significance of fanlike throw-off has never been discussed and numerically modeled, so it is unclear to what extent it contributes to the resultant aerodynamic force generated while flying, and whether it is a separate aerodynamic phenomenon or simply a wing movement that enhances circulation, such as the swinging-edge mechanism. However, a strong flow occurs in consequence with the throw-off, and, thus, flying specimens are presumably able to use the throw-off to generate useful force. The throw-off is probably more important for insects with the most intensive pronation and supination, principally the dipterous flies.

In the crane fly, the process of vortex formation during upstroke looks somewhat different (fig. 2.12G, H). The vortex under the abdomen has never been observed, and presumably does not exist, indicating that the stopping tubes of the crane fly do not meet and merge under the body. This is probably because of the very narrow base of crane fly wings, which are affixed to the body by means of thin stalks. The small stalk diameter does not allow the wings to accelerate air actively during strokes and thereby generate useful forces. Therefore, the proximal parts of the stopping tubes do not spread farther toward each other after they reach the stalks at the beginning of the second half-stroke, but instead remain closed to the proximal parts of the wing plate (fig. 2.12G). As a result, the ring remains open for the main part of the upstroke. During this period, the stopping tubes are strongly curved, so that each tube can be cut twice by a horizontal plane in positions 8–10. Two vortices are seen at each body side, one near the wing tip and the other at the proximal part of the wing, rotating in opposite directions to each other (Grodnitsky and Morozov 1993). In a tethered crane fly, the vortex ring closes quite late, when the wings approach each other at the end of the stroke cycle (fig. 2.12H). Then the entire ring can be found over the insect, so that the ventral vortex is situated not under, but over the body.

It is difficult to determine to what extent this result reflects the structure of the aerodynamic wake of the crane fly in natural flight. In a natural situation, it is possible that the ring closes over the body, as determined experimentally, owing to the specific shape of the crane fly wings. However, in several cases the ring was observed to close above the body of insects with broad wing hinges, for example, of noctuid moths (Grodnitsky and Morozov 1993). With respect to these findings, it is easier to believe that transfer of the ventral vortex onto the dorsal side results instead from a stroke with incomplete amplitude, when the wings open to the positional angle of 90–100° or less. Possibly the stopping tubes merge under the abdomen during the normal flight of crane flies, as they do in other insects.

Dickinson and Götz (1996) described another behavioral modification of the ventral tubes in *Drosophila*. They demonstrated that during upstroke the tubes fuse neither under nor over the body, but remain attached to the fly body until the stroke cycle is finished. In the course of supination, the ventral tubes detach from the tip of the abdomen, fuse, and thus form a self-closed, strongly curved vortex loop. Note, however, that the flow passing through the loop is structured; thermoanemometry procedure indicates three flow corridors inside it, all parallel to the longitudinal body axis, one located above the insect and two at the sides. The three corridors meet, unite, and form a single, round-sectioned flow 5 mm behind the stroke plane (Lehmann 1994; Dickinson and Götz 1996).

2.5. Vorticity at Posteromotorism

Beetles have a peculiar mode of posteromotoric, functionally four-winged flight and derived kinematic patterns. According to Schneider (1975) and the data of Grodnitsky and Morozov (1994), the Scarabaeida possess three main types of flight: (1) the elytra do not open and do not contribute to forces necessary for flight (this state is morphologic two-wingedness); (2) the elytra open and perform strokes at comparatively low (30–50°) amplitude (this is functional four-wingedness with leading hindwings); (3) the elytra open and remain immobile throughout the entire flight (functional two-wingedness with uncoupled wings). In the first case, the beetles can be assumed to have flight aerodynamics similar to other morphologically two-winged insects. In this section, data on the vortex wakes of beetles of the second and third types are presented. The soldier beetle *Cantharis* sp. and the tiger beetle *Cicindela restricta* represent these respective flight types.

During the period of each wing stroke cycle, generation and development of vortex structures are observed. When the wings move downward from the upper point of their trajectory at the beginning of the tiger beetle wing stroke, a dorsal vortex emerges between them (fig. 2.16*A*), indicating that circulation flow dominates around the wings. This vortex constitutes a section across the U-shaped vortex bubble located over the insect body and closed to the wing tips by its ends. As the downstroke proceeds, the diameter of the dorsal vortex increases (fig. 2.16*B, D*). Judging from obtained photographs, the rotational speed increases as well. During the first half-stroke, a gradual backward displacement of the dorsal vortex occurs, so that the vortex is situated over the tip of the beetle abdomen during supination and reversal of wing motion. This displacement is characteristic of all insects that do not possess secondarily narrowed wing hinges, such as lacewings (see sec. 2.6) or crane flies (see sec. 2.4).

The wings of the tiger beetles come close to each other under the abdomen during the course of supination, owing to the high stroke amplitude. The ends of the U-shaped bubble closed to the wing tips approach each other as well. In the bottom of the stroke or shortly after, the ends of the U-bubble meet, so that a complete self-closed vortex ring appears. At this moment, a ventral vortex emerges on the flat sections of the flow, rotating in the opposite direction to the dorsal vortex (fig. 2.17), resulting from closure of the U-shaped bubble under the insect body.

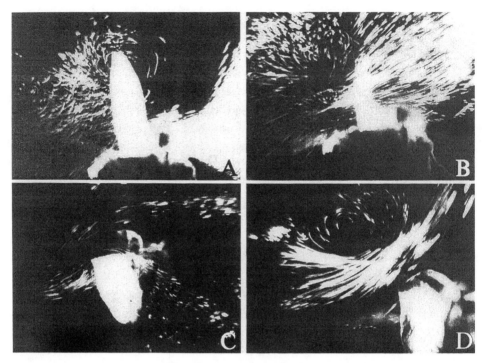

Figure 2.16. Dorsal vortex above the body of tiger beetle at sequential moments (*A–D*) during downstroke. Insect flying to the right.

During the entire second half-stroke, the lower vortex is located under the beetle abdomen. Its diameter is noticeably less than that of the dorsal vortex (fig. 2.17). The caudal displacement of the dorsal vortex in the first half-stroke does not allow room for the formation of a separate ring during upstroke. Accordingly, during part of the upstroke (at least its lower half), the wings move inside the vortex ring generated in the downstroke. The ring travels backward as pronation starts, and a new stroke cycle begins, followed by the corresponding formation of new vortex structures (fig. 2.16*A*). The formation of vortices near the elytra of the tiger beetle is not observed during the stroke cycle: airflow around the elytra is smooth and does not separate from either the trailing or leading edges (fig. 2.16*C, D*).

The general schemes of tiger beetle and soldier beetle vortex wakes do not differ substantially: during each stroke, dorsal and ventral vortices are formed over and under the body of *Cantharis* (fig. 2.18) in the same manner as in *Cicindela*. This reflects the similarity of wake structure and suggests the formation of a single ring per stroke. Thus, based on the available data, the

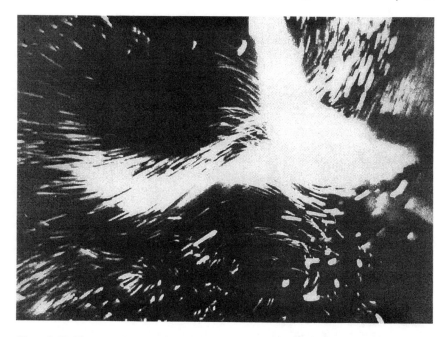

Figure 2.17. Vortices above and under the body of tiger beetle during upstroke. Insect flying to the right.

aerodynamic wake of flying beetles is represented by a sequence of unlinked parallel rings, as in other insects studied in this respect. Reduced strokes of the elytra do not change the schematic diagram of the wake.

The pictures of flow around the elytra of tiger beetles indicate that they do not create their own vortex structures. This is possibly owing to their immobility, or to the suction effect of flapping hindwings, which causes rarefaction and thus prevents formation of separate vortex structures behind the elytra. However, a starting vortex behind the elytra at takeoff and a stopping vortex at landing must be produced. Then, circulation flow can be established around the elytra, accompanied by two tip vortices that come off the elytra tips and are accumulated by the nearest vortex ring. If this occurs, the immobile elytra of tiger beetles perform analogously to an aircraft lifting surface (see fig. 2.4C). In this case, the lift created by the elytra strongly depends on flight speed: the higher the speed, the more effective the lifting surface; in hovering it is useless.

The latter assumption agrees well with observations of the flight of beetles in nature. Tiger beetles inhabit warm, open plots. The use of immobile stretched elytra as lifting surfaces required the evolutionary development

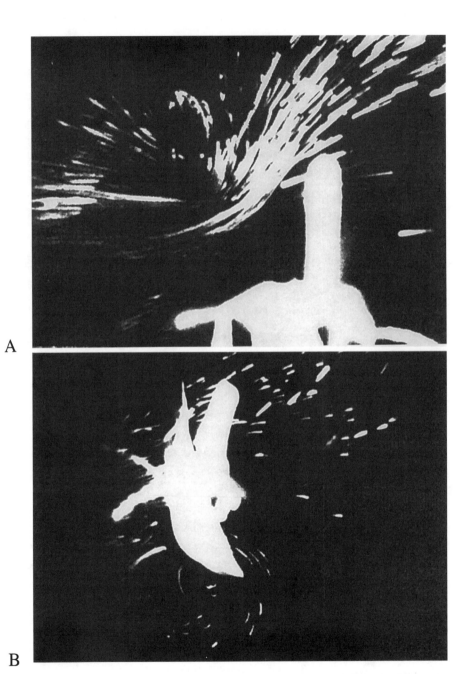

Figure 2.18. Vortices by soldier beetle during downstroke (*A*) and upstroke (*B*). Insect flying to the right.

of a flight behavior so extremely specific that identification of this beetle genus is possible from a great distance. These beetles take off instantly and their flight is swift and, in the majority of cases, straight. Deceleration before landing is very rapid and imperceptible to the naked eye. Hovering or similar flight modes have never been observed and are most likely impossible. Moreover, tiger beetles have obtained an accompanying set of morphologic features, which justify the separation of *Cicindela* and related genera into a family (Shwanvich 1949; Borror et al. 1992).

In contrast to tiger beetles, soldier beetles inhabit dense herbaceous vegetation. Their takeoff is comparatively slow. While flying, the soldier beetles often change direction, performing turns at 180° and greater. "Landing" onto herb stems requires considerable accuracy and, above all, the capacity for slow horizontal or hovering flight. I have observed rather prolonged (up to 15–30 s) natural hovering of soldier beetles. This suggests that their elytra generate lift not only in fast flight (i.e., owing to intensive work of the hindwings), but also independently, inducing their own flow during each stroke. Based on this finding, one should anticipate that in the beginning of each stroke, each elytron generates its own starting vortex. However, the vortex wake of soldier beetles contains no structures that could arise from the development of such vortices. It can therefore be assumed that during each stroke cycle, the starting vortices of the elytra merge with the more powerful upper U-shaped bubble of the hindwings. Nevertheless, not even vestiges of vortices have been observed near the trailing edges of elytra in the beginning of the stroke of soldier beetles. Most likely, formation of the starting tubes is handicapped by a suction effect from the vortices of the hindwings. This cannot prevent the establishment of circulation flow around the elytra, because development of the starting vortex is not the cause, but rather a consequence of the growth of circulation. As a result, the impulse of the generated vortex equals the sum of circulations around the wings and elytra.

The presented data and considerations do not coincide with the early opinion of Pringle (1963), who believed the elytra to be totally passive organs that oscillate only under the effect of vibration resulting from the use of the metathoracic musculature. The data instead confirm the viewpoint that the elytra of beetles act solely as generators of lift (Nachtigall 1964; Schneider and Hermes 1976). Owing to their restricted mobility and inability to pronate and supinate, elytra do not generate thrust.

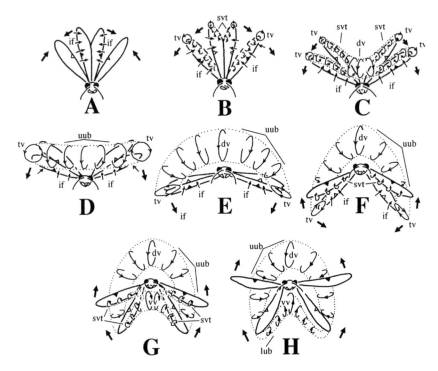

Figure 2.19. Spatial structure of the flow around green lacewing during one wing-beat cycle; sequential stroke phases (frontal views). (*A*) Irrotational flow (if) about the leading edge. (*B*) Independent generation of single vortex tubes (svt) above each wing; tv, tip vortices. (*C*) Start of coalescence of the vortex tubes; dv, dorsal vortex. (*D*) Formation of the upper U-shaped bubble (uub). (*E*) Growth of the uub. (*F*) Generation of stopping vortex tubes below supinating forewings. (*G*) Generation of stopping vortex tubes below the hindwings and the beginning of coalescence between single vortex tubes; vv, ventral vortex. (*H*) formation of the lower U-shaped bubble (lub) and birth of a single vortex ring. Large arrows, movement of the wings; small arrows, movement of air.

2.6. Vorticity in the Flight of Anteromotoric Functionally Four-Winged Insects

The green lacewing *Chrysopa dasyptera* serves as a representative functionally four-winged insect. In the beginning of each stroke, movement of the wings causes nonseparating airflow between their dorsal surfaces (figs. 2.19*A* and 2.20*A*): the flow goes smoothly around the leading edge and deflects to the wing tips. Consequently, as the trailing edges begin to separate, a difference of air speeds already exists near the upper and lower sides of each wing. This

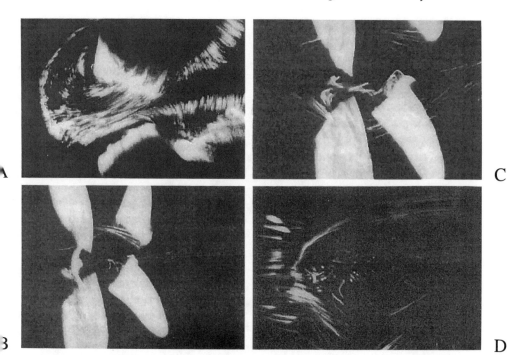

Figure 2.20. Flow around the wings of green lacewing at the beginning of the down-stroke; sequential phases, viewed from above: (*A*) the forewings pronate while the hindwings (not illuminated) still move upward to meet each other; (*B*) both wing pairs in early downstroke; (*C*) the same slightly later; (*D*) same, after both fore- and hindwings have gone off the beam plane. Insect flying to the left.

facilitates quick formation of starting vortex tubes over the trailing edges of the wings (figs. 2.19*B* and 2.20*B*). Each tube is closed to the wing tip by one of its ends, and to the proximal part of the trailing edge by the other; its section near the wing is presented by a tip vortex (fig. 2.21*B*). While the stroke cycle continues, the tubes of all four wings accumulate the vorticity coming from the wings and increase in size and speed of rotation (fig. 2.20*C, D*). As flow speed increases around the wings, the tubes elongate, so that their proximal ends come closer to the body and finally unite with the tubes on the opposite side of the body. Owing to mutual attraction, the tubes of the fore- and hindwing pairs simultaneously merge and form a common U-shaped vortex bubble situated over the insect body (fig. 2.19*C, D*). The section of this U-shaped bubble by a sagittal plane is the dorsal vortex, clearly visible in many photographs (fig. 2.21*A*). The coalescence of the tubes occurs comparatively early, before the positional angle of the wings reaches 30°.

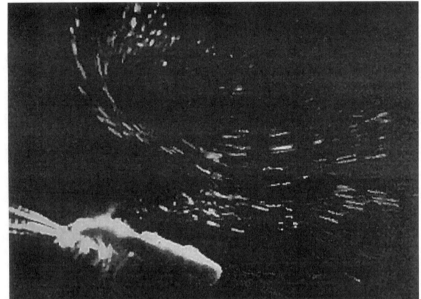

A

B

Figure 2.21. Flow around the wings and body of green lacewing during downstroke: (*A*) dorsal vortex (lateral view); (*B*) tip vortex on right forewing (inclined view). Insect flying to the left.

Throughout the entire downstroke, the U-shaped bubble is situated directly over the body and wings (fig. 2.19*E*). When the forewings reach the lowest point of their trajectory and start to supinate, the U-bubble passes onto the tips of the hindwings, which continue to move downward and therefore experience larger negative pressure than the forewings. A similar phenomenon was described in anteromotoric functionally four-winged caddisflies (Ivanov 1990).

The beginning of the second half-stroke of the forewings is accompanied by the formation of stopping tubes, which rotate in the opposite direction to the upper U-shaped bubble (fig. 2.19*F*). The stopping tubes of the left and right wings unite in the same manner as during the first half of the stroke cycle.

Supination of the hindwings results in the formation of two more stopping tubes (fig. 2.19*G*). Subsequently all the vortices below the insect merge (fig. 2.19*G*). This fusion is indicated by the presence of a single vortex under the abdomen of the insect (fig. 2.22). The upper and lower tubes unite into a single vortex ring via the wing tips (fig. 2.19*H*). As in other insects (see sec. 2.4), the ring shape noticeably differs from a toroid, so that frontal views reveal only the upper two-thirds of the ring. The lower part remains shaded, since it is situated under the abdomen, behind the light plane crossing the wing tips (fig. 2.23).

These results indicate that each wing of an anteromotoric functionally four-winged insect works as a separate generator of vorticity: at the points of movement reversal, vortices independently come off the fore- and the hindwing pair. In the course of a stroke, the vortex tubes of different wings merge and form a single vortex ring, which loses connection with the wings and travels away from the insect in the beginning of the next stroke cycle. Each stroke is therefore followed by the generation of one vortex ring.

2.7. The Problem of Flow Separation from the Leading Edge

According to Weis-Fogh (1973, 1975), pronation of the wings in the very beginning of the stroke cycle acts as a specific mechanism for force generation. Based on data from high-speed cinematography of free flight of the tiny wasp *Encarsia formosa*, Weis-Fogh described the process as a booklike opening of the wings, initially touching each other by their upper sides. He termed this movement *fling*. Fling was later observed in additional species (Cooter and Baker 1977; Antonova et al. 1981; Brodsky 1991; Brackenbury 1992); for more details of the fling mechanism see Ellington (1984, 1995).

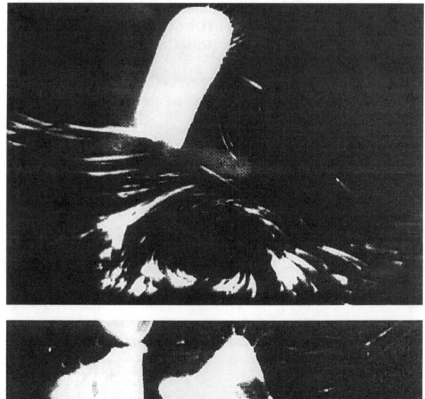

A

B

Figure 2.22. Flow around the wings and body of green lacewing during upstroke: (*A*) vortex under a hindwing (lateral view); (*B*) same (seen from above). Insect flying to the left.

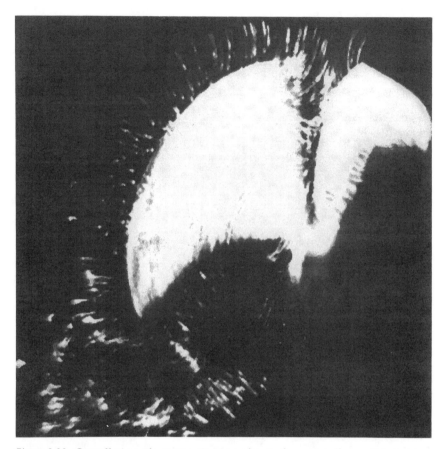

Figure 2.23. Overall view of a vortex ring just after its formation (for spatial structure see fig. 2.21*H*). Viewed from the front, light plane position 12 (see fig. 2.7).

Throughout the last two decades, fling has been the focus of most active research and debate in the field of insect flight aeromechanics. In Weis-Fogh's interpretation, a useful aerodynamic effect is achieved when air enters the space between the wings, accelerates, and creates rarefaction. Fling can generate lift even in an ideal fluid, without any accompanying vortex formation, thus constituting a fundamentally novel case for classical hydroaeromechanics (Lighthill 1973, 1978). Nevertheless, the useful effect of the fling movement could be increased if flow separated from the leading edges and formed a corresponding vortex (Lighthill 1973, 1978), causing an increase in air velocity over the upper surface of the wing and hence producing additional lift. Assessment of this hypothesis by the mathematical modeling of

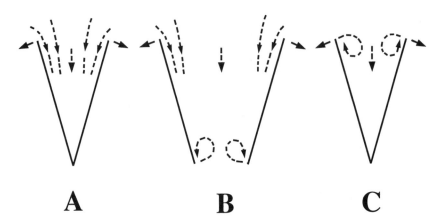

Figure 2.24. Two-dimensional flow near the wings in the beginning of stroke cycle during the fling: living insect (*A, B*) and mechanical model (*C*) (after Maxworthy 1979; Wilkin 1985). (*A*) Smooth flow near the leading edge; (*B*) stalling on the trailing edge and the formation of starting vortex; (*C*) stalling on the leading edge and the formation of leading-edge vortex bubble. Seen from above.

two-dimensional flow around a pair of opening wings showed that with flow separation from the wing leading edge, a significant complementary force could indeed be generated (Belotserkovsky et al. 1974a, 1974b; Belotserkovsky and Nisht 1978; Edwards and Cheng 1982). Experiments on visualization of flow around flapping mechanical models confirmed the presence of the separation that caused the generation of the specific vortex cord near the leading edge (Maxworthy 1979; Saharon and Luttges 1987, 1988; Ellington et al. 1996). Given these results, the separation has been incorporated into the majority of aerodynamic models of insect flapping flight (Maxworthy 1981; Ellington 1984; Wilkin 1985; Spedding and Maxworthy 1986; Brodsky 1988; Sunada et al. 1993a). Contrary to this conventional opinion, the information presented on flow around flapping insects shows that after the beginning of the stroke cycle, during the fling movement, airflow passes by the leading edges smoothly (fig. 2.20*A*). Flow separation and subsequent growth of vorticity occur near the trailing edges not during, but after fling, as the trailing edges depart from each other. The vortices generated therewith cannot be interpreted as anything other than starting vortices (fig. 2.24).

The hypothesis of flow separation from the leading edge primarily originated from an analogy with man-made aircraft. Opportunities for the technical application of separated flows on the leading edges of thin wings have been discussed for several decades. Separated flow generation was suggested to increase the lift-to-drag ratio of the B-70, Concorde, and Tu-144 (Batchelor 1967; Ashley 1974; Belotserkovsky and Nisht 1978), and of vertical and short takeoff and landing aircraft (Rossow 1978). Methods of utilizing the separation generated by oscillating airfoils or by pulsations of flow around static airfoils continue to be actively investigated in aircraft engineering (Maresca et al., 1979; Favier et al. 1982; McCroskey 1982; Robinson and Luttges 1983; Ashworth and Luttges 1986; Ashworth et al. 1988; Schreck and Luttges 1988; Freymuth 1990; Ohmi et al. 1990, 1991; Gurgey and Thiele 1991; Gursul and Ho 1992).

Application of the flow separation principle to the description of flight of living beings is more problematic. Vortex formation undoubtedly occurs at the leading edges of motionless artificial models of insect wings placed in a wind tunnel (Vogel and Feder 1966; Vogel 1967b), or performing strokes in a fluid (Maxworthy 1979, 1981; Savage et al. 1979; Spedding and Maxworthy 1986; Dickinson and Götz 1993; Sunada et al. 1993a; Dickinson 1994) or in air (van den Berg and Ellington 1997a, 1997b). In addition, data exist that demonstrate nonseparated flow around a motionless model of the wing of the flower fly *Syrphus balteatus* (Rees 1975). Furthermore, flow visualization reveals that actual detached insect wings also lack separation from the leading edge (Vogel 1967b). A similar conclusion can be derived from the shape of aerodynamic polar curves that correlate dynamic characteristics of a wing with its angle of attack. Polar curves, as a rule, do not indicate a dramatic reduction of lift after the wing achieves supercritical values of the angle of attack. This suggests that the increase of this angle does not result in flow separation from the wings of most different insects: pomace fly *Drosophila virilis* (Vogel 1967b), dragonflies *Aeshna* sp. (Newman et al. 1977) and *Sympetrum sanguineum* (Wakeling and Ellington 1997a), damselfly *Calopteryx splendens* (Rudolph 1976a; Wakeling and Ellington 1997a), crane fly *Tipula oleracea* (Nachtigall 1977), and various butterflies and moths (Nachtigall 1967). Separation has been shown only on the hindwing of the desert locust *Schistocerca gregaria* immovably fixed in a wind tunnel, whereas the forewing of this species did not indicate the separation (Jensen 1956).

From the perspective of conventional steady aerodynamics, this finding is not surprising. Polar curves indicate an evident crisis of flow around a still wing at Reynolds numbers (Re) 10^6 and larger, whereas at $Re = 4 \times 10^4$ the curve already displays a smooth shape, and at Re from 10 to 10^3, flow separation is absent (Getzan et al. 1995; Shimoyama et al. 1995). One can argue that the steady situation cannot be related to insect flight, which is in principle an unsteady phenomenon, because the velocity of airflow around insect wings constantly changes in direction and magnitude. Nevertheless, note that in steady flow under sufficiently large *Re*, separation does not appear immediately after the wing achieves a supercritical angle of attack. The process is inertial, and time is necessary for the development of separation on the leading edge. The wingbeat cycle in insects lasts 10^{-1}–10^{-2} s and less. Therefore, separation of flow from their wings is even less predictable than in a steady situation.

Visualization experiments with tethered flying caddisflies (Ivanov 1990) and free-flying *Heliconius* butterflies (Grodnitsky and Dudley 1996) also fail to support the separation hypothesis. Dickinson and Götz (1996), having visualized tethered-flying *Drosophila* and having found no separation from the leading edge, suggested that the vortex filament is too thin to be detected by their method. In this case, with *Re* about 100 for *Drosophila*, it should be less than 10 for such a small undetectable vortex. Vorticity generation at these *Re* values is highly questionable, even more so because the time to create the vortex is quite short (about 2×10^{-3} s or less, since the wingbeat frequency of *Drosophila* in the experiments of Dickinson and Götz [1996] was close to 200 Hz).

Convincing data in support of the separation hypothesis were recently presented, arguing that flow separates from the wings of live flapping insects after the fling movement is complete, and the wings undergo downward translation (Ellington 1995; Ellington et al. 1996; Willmott et al. 1997). Flow separating from leading edges is clearly seen at high (1.8 m/s and greater) flight speeds, whereas it is much less evident at lower speeds. The opinion of Ellington, Willmott, and their coworkers is that at low flight speeds, the leading-edge vortex bubble is too small to be seen in smoke visualization photographs. They also state that the bubble, in principle, cannot be registered under light plane illumination, because the strong axial (from hinge to tip) flow that occurs during the downstroke imparts a conical spiral shape to the leading-edge bubble. As a matter of fact, any vortex that develops during a stroke of flapping wings is spiral. The tip vortices are spiral as well (owing to the progressive downward wing movement rather than the axial flow), but they still can be detected easily in photographs (fig. 2.21*B*), since the light plane width is not zero. Probably, vorticity is absent on the leading edge of flying insects in hovering and slow forward flight, showing a difference between live specimens and artificial models.

This difference has been attributed to two properties of insectan wings. The first suggestion reflects the fact that the surface and margins of insect wings are covered with microscopic structures (hairs, spines, scales, etc.) (Vogel 1967b). An approximate calculation according to Prandtl's formula (1952, cited in Vogel 1962) shows that superficial microstructures do not project outside the boundary layer and hence do not interact with free flow (Grodnitsky et al. 1988; see also Bocharova-Messner 1979a). Note, however, that Prandtl's boundary-layer equation is applicable to conditions of stationary flow, when the speed and direction of flow are unchanged in time. This condition is unfulfilled in insect flight. Nonetheless, although formulas for calculation of the boundary layer on a flapping wing do not currently exist, the thickness of the layer must be the same as or greater than on a stationary lifting surface of the same shape, because the development of flow and establishment of minimum boundary-layer thickness can demand a greater time period than the stroke cycle allows.

Thus, according to rough quantitative data, microstructures do not interact with free flow. However, they still can somehow affect the boundary layer and prevent separation (Vogel 1962). The mechanisms of such an effect can differ. On the wing surface, microscopic appendages are assembled in particular comblike structures that can cut the boundary layer into microflows (Bocharova-Messner 1979a, 1982) and make them turbulent (Zakharenkov et al. 1975). It is also possible that spines and scales simply increase the total wing surface (and correspondingly the surface of the contact between the wing and boundary layer), thereby resulting in a decrease in the flow inside the layer and preventing separation. If the microstructures are able to prevent separation, they probably do so only in a limited range of flow speeds near the wings. This is suggested by data on flow separation from the wings of a large hawkmoth in fast flight (Ellington 1995; Ellington et al. 1996; Willmott et al. 1997). The trailing edge provides another counterargument. This region normally bears much more highly developed microstructures than the leading edge, yet they do not prevent flow separation. It is admittedly conceivable that separation from this part of the wing is induced by the difference of pressure that originates after fling but is absent in the very beginning of the stroke cycle.

The second reason suggested for nonseparating flow around the leading wing edge in a limited range of flow speeds is the flexibility of insect wings: they are elastic structures and bend toward the opposing wind during fling (the flexible fling modification is designated by the term *peel*) (Ellington 1984). This viewpoint also encounters contradictions, more specifically the lack of vortex formation near the rigid elytra of bugs and beetles. My personal opinion is that separation of flow is not produced by the leading edges of insect wings primarily because of the small size of the wings and the low *Re* at which they operate. From this perspective, it is not unusual that flow separation has been reported only for the locust and the hawkmoth, which are among the largest species of insects.

Summarizing these considerations, separation of airflow from the leading edges of flapping wings can be seen in experiments with mechanical models, but is absent in live individuals in hovering and slow forward flight, although such separation could potentially contribute useful forces. According to the data of Ellington and coworkers (Ellington 1995; Ellington et al. 1996; Willmott et al. 1997), the separation appears instead at large *Re,* when the insect is big enough (as *Manduca sexta*) and flight velocity is high (1.8 m/s and greater).

For medium-sized and tiny insects in slow and hovering flight, flow separation remains problematic and probably does not exist. Furthermore, it has been shown repeatedly that the aerodynamic force generated via conventional mechanisms is insufficient to explain the nature of insect flight (Ellington 1995; Dickinson and Götz 1996; Ellington et al. 1996). Until the present time, leading flow separation was the single phenomenon capable of acceptably uniting the mathematical modeling of insect flight with empirical data. The flow separation is of essential importance to the theory of hovering — ironically, the specific situation in which it has never been observed. Because such separation appears to be rarer than the amount of data to be interpreted, alternative unsteady mechanisms must be found. Such a mechanism could consist, for example, in the interaction of the wings with the starting vortex, clearly visible in any insect, although the effect of its presence close to the wings has not yet been assessed.

2.8.   The Problem of Force Generation

A flapping wing constantly operates in near-to-start conditions. This means that within any stroke it has no time to generate the maximum circulation possible at its size, shape, and initial speed. The delay in the growth of circulation is called the *Wagner's effect*. It has been suggested that insects overcome this effect by interacting with their own vortex wake, whether or not they have generated the leading-edge separation bubble (Savchenko 1971; Lan 1979; Ellington et al. 1996).

In contrast to man-made aircraft, the vortices generated by the wings of insects are not discarded after takeoff, but are continuously renewed within each second of flight. Vortex tubes are zones of reduced pressure. Located close to the insect, they create rarefaction and, consequently, complementary forces. Hence, natural selection may have contributed to the development of adaptations that allow interactions between the wings and vortices in order to minimize negative influence on the tubes from air viscosity to recover part of the energy the wake obtained from the wings. Uldrick (1968) initially offered this hypothesis almost three decades ago. It is now possible to suggest a version of it relevant to insect flight.

First, insects could have developed distinct mechanisms that cause earlier merging of the left and right starting tubes. As previously shown for lepidopterans, full clap over the body and subsequent separation of the wings leads to the early formation of an upper U-shaped bubble. Consequently, Weis-Fogh's fling, most distinctly expressed in lepidopterans, can act as an adaptation to control the near vortex wake shape.

Second, the formation of the independent starting tubes behind each wing in anteromotoric functionally four-winged insects at the beginning of each stroke cycle results in the tubes decelerating each other prior to merging. Accordingly, the aerodynamic properties of the flying system could be improved by the wings coupling into two flapping planes and generating a single tube behind each couple.

Third, throughout the entire downstroke the upper U-shaped vortex bubble is closed to the tips of flapping wings, thereby creating rarefaction over the flying specimen. It can be assumed that the insect derives some energetic gain if it prolongs any positive influence of this bubble. Therefore, supination still continues after the beginning of the upstroke. As a result, stopping tubes do not form until the supination is finished. The "aerodynamic" downstroke, therefore, lasts longer than the "kinematic" downstroke (see sec. 2.1).

Fourth, during upstroke the stopping tubes are located under the wings and most likely create additional negative aerodynamic force. From the viewpoint of natural selection, it would be useful to diminish the negative influence of the stopping tubes. This can be achieved by shortening the duration of the wings' upward movement, decreasing their angle of attack, and delaying merging of the tubes under the abdomen and the following formation of the lower U-shaped vortex bubble.

The shorter duration and low angle of attack during the upstroke are well known (see sec. 2.1). The delay of formation of the lower U-shaped bubble is caused by an incomplete stroke in the bottom of the trajectory, since the wing positional angle normally does not exceed 120–150°. For this reason, the proximal ends of the tubes appear to be parted from each other, and they need additional time to meet and merge. Incomplete downstroke is a functional way to prevent stopping bubbles from connecting. Dipterous flies are the most advanced insects in this respect, having narrow wing hinges, which increase the distance between the proximal ends of the stopping tubes that close against the ano-jugal zones of the wings. Accordingly, the lower U-shaped bubble remains open throughout the main part of upstroke and sometimes even closes above the body (see sec. 2.4).

The discussion of problems involving analytical description of the aeromechanics of insect flight is beyond my professional ability. However, it is necessary to note several points that could be important for the numerical description of the flapping flight of living beings. Recently the opinion predominates that the quasi-steady approach to calculation of the wing force parameters through the representation of the stroke cycle as a series of steady-state conditions is an unrealistic simplification of animal flight, which is instead predominantly based on unsteady effects, so that an appropriate description of the natural phenomenon cannot be achieved in this way (Ellington 1984, 1995; Dudley and Ellington 1990b; Spedding 1992, 1993; Ellington et al. 1996; Willmott et al. 1997). More reliable vortex theories derive the force necessary for flight from the momentum that the wings transfer to air (Ellington 1977, 1980, 1984; Rayner 1979a, 1979b, 1980; Azuma and Watanabe 1988; Azuma 1992). Nevertheless, comparison of available vortex theories with vortex impulse measurement data shows that the vortex wake energy is approximately twofold less than that demanded for free flight (Spedding et al. 1984; Spedding 1986, 1992). Based on these considerations, the impulse shortage can be connected with the interaction of wings with the leading-edge vortex (whereby it is generated) and/or with the starting vortices. Another additional force can be created by pronation and supination.

As the most simplified and generalized representation, a stroke cycle can be imagined as a succession of different aerodynamic mechanisms. It begins with a fling movement (in large insects possibly accompanied by flow separation from the leading edges). Fling is followed by translation of the wings, formation of starting vortices over the trailing edges, and subsequent growth of circulation and the corresponding vertical force. Both the circulation and the starting vortex can be enhanced by the swinging-edge mechanism. During downstroke, the starting vortex locates near the wing, influencing the entire flow field and accelerating air above the insect. Supination is accompanied by fanlike airflow coming off the upper side of the wing. Simultaneously, upstroke begins and the stopping vortex is formed, always of a smaller size than the starting bubble. After a while, the vortices over and under the wing unite into a vortex ring. The ring departs from the wings at the end of the upstroke under the effect of fanlike throw-off. The varying nature of the processes that accompany the progress of the wingbeat cycle suggests that a different algorithm should be used to describe each phase of the stroke cycle.

2.9. Body Motion during Stroke Cycle

2.9.1. Motion of the Pterothorax

The aerodynamic significance of the phenomena that occur in the air within each stroke phase can be assessed according to the character of motion undergone by the insect body during the stroke phases. To examine this problem, the periodic vertical and horizontal oscillations that follow each stroke cycle were studied in free-flying butterflies of several North Asiatic species of the family Nymphalidae (Grodnitsky 1992a, 1993). Based on measurements of pterothoracic oscillations, it was possible to ascertain the distribution of lift throughout an averaged stroke cycle.

According to cinematographic data, a stroke cycle is divided into five phases (fig. 2.25). Before the beginning of each stroke, the wings tightly adjoin each other with their upper sides over the insect body. Immediately after beginning, the fling movement takes place (phase 1), followed by phase 2 (translation), when the wings move downward with an approximately constant angle of attack. (Owing to a comparatively small frequency of framing, the swinging-edge movement was not separated into a single phase.) The stroke continues in phase 3, which coincides with supination. Phase 4 is the upstroke followed by pronation (phase 5). During this final stage the wings approach and touch each other, in a motion named *clap* (Weis-Fogh 1973). Clap and fling compose two successive phases of pronation.

Frame-by-frame analysis of cinematographic films provided an opportunity to plot histograms that reflect the within-stroke distribution of pterothorax motions (fig. 2.26A, B). It was found that in horizontal flight, butterflies descended during the absolute majority of recorded claps and flings. Climbing was observed in phases 2, 3, and 4. The distance traveled by the body exceeded its mean value in phases 1, 2, and 5, and was less than the mean value during supination and upstroke.

Measurement of the vertical and horizontal displacements of the butterfly

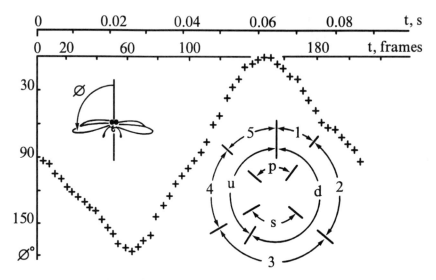

Figure 2.25. Dependence of wing positional angle (∅, degrees) on time (t) during stroke cycle of *Leptidea sinapis* L., according to the data of frame-by-frame analysis of a high-speed cinematographic film. The angular diagram indicates interrelationships between the main wing movements: downstroke (d), upstroke (u), pronation (p), supination (s), and their correspondence to the distinguished stroke phases (1–5).

bodies during horizontal flight enabled a graph to be plotted that reflects the character of vertical motion of the center of mass in each stroke phase. At first approximation, the curve has the shape of a sinusoid (fig. 2.26C). The force that induces sinusoidal oscillations is proportional to the second derivative $\sin''x = -\sin x$. Consequently, within-stroke lift distribution is approximately described by an antiphase sinusoid, which means that maximum lift occurs during the downstroke. The maximum value of thrust cannot be correlated to any stroke phase, because thrust is opposed by drag, which is continuously changing in conjunction with flight speed pulsations. By contrast, in vertical oscillations the generated lift is counterbalanced by the weight of the insect, which is constant and thus allows assumptions to be made about lift dynamics during stroke cycle.

A similar pattern of lift distribution throughout the stroke cycle was obtained from studies of the flight of the stable fly and dragonfly, tethered to highly sensitive strain gauges that measured the vertical and horizontal forces acting on the body during flight (Buckholz 1981; Somps and Luttges 1985). According to these studies, lift is close to zero at the beginning of the stroke. Then lift increases, achieves its extreme value in the second half of

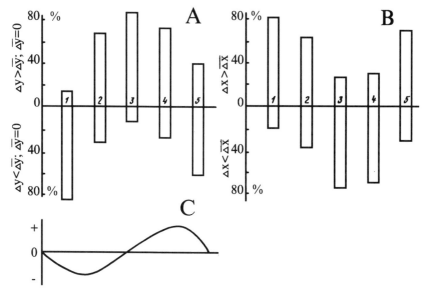

Figure 2.26. Within-stroke distribution of vertical (*A*) and horizontal (*B*) displacements of butterfly pterothorax during slow forward flight and corresponding hypothetical reconstruction of vertical oscillations of pterothorax (*C*). The data are summarized for three species: *Aglais urticae, Inachis io,* and *Polygonia c-album.* Δy, Δx = vertical and horizontal displacements that take place during a single stroke phase (for indication of the phases, see fig. 2.25); $\overline{\Delta y}$, $\overline{\Delta x}$ = mean estimates of these values.

downstroke, begins to decay during supination, and assumes negative values throughout upstroke. By contrast, thrust is maximal in the middle of upstroke (phase 4). During phases 1–3 and 5, thrust is less than the drag experienced by the insect during flight.

Similarity between the hypothetical distribution of lift within the stroke cycle in free-flying butterflies and the measured lift distribution in the tethered fly suggests that aerodynamic force generation in these insects is realized according to the same principles. One can also conclude that experimental tethering of insects does not change these principles excessively, and that tethered flight therefore reflects the primary phenomena that accompany natural flapping flight.

The reported data enable one to appreciate, at first approximation, how some of the processes in the air influence the overall balance of useful and parasitic forces that affect a flying insect. In all likelihood, fling, initially described as a qualitatively novel aerodynamic adaptation to flapping flight (Lighthill 1973; Weis-Fogh 1973), indeed favors fast growth of lift. On the other hand, the chalcid wasp *E. formosa* counterbalances its weight after, rather than during fling (since the descending part of the flight path is terminated within phase 2), a finding that contradicts the results of Weis-Fogh's (1973) investigation of free flight in this species. Consequently, fling favors quick growth of lift and constitutes a stage preparatory to the main part of downstroke, when the wings create the force necessary for the insect to stay aloft.

Bocharova-Messner and Aksiuk (1981) and other researchers (Betts and Wootton 1988; Brodsky 1991; Willmott et al. 1997) believe that thrust can be partially generated owing to jet effects. Indeed, as can be seen from the presented data, phase 5 coincides with the interval of positive thrust values. Nevertheless, this assumption, in general, appears doubtful for insects. Examination of the structure and dynamics of the flow around flying insects shows that the wings throw a volume of air backward in a fanlike manner during pronation (corresponding to phases 5 and 1). However, the flow picture observed during this stroke component differs fundamentally from predictions implicit in the jet hypothesis (see sec. 2.4).

2.9.2. *Motion of the Abdomen*

Straight flight is accompanied by periodical deflections of the butterfly's abdomen from the longitudinal axis of the thorax. During the principal part of the stroke cycle, the abdomen is kept in line with the pterothorax and deflects upward during supination, when the wings achieve a positional angle of 130–160°. The amplitude of these deflections ranges between 10 and 30° in the species of butterflies investigated. A change of flight direction was repeatedly observed in cases in which the abdomen remained in line with the pterothorax at the end of downstroke.

Throughout phase 3, the body of the butterfly acquired a more vertical position, and after this the insect began to gain height. The upward turn was quicker when the abdomen, having declined upward earlier, began to move in the same direction as the wings at the end of downstroke (fig. 2.27). Stability of general flight direction demands that the point of force application is located on the longitudinal axis of the body of a flying specimen, in front of the center of mass, which resides in the proximal part of the abdomen in different insects (Magnan 1934; Ellington 1984; Srygley and Dudley 1993). Therefore, when the wings create maximal lift at the end of the first half-stroke, a torque arises that tends to turn the insect upward about the center

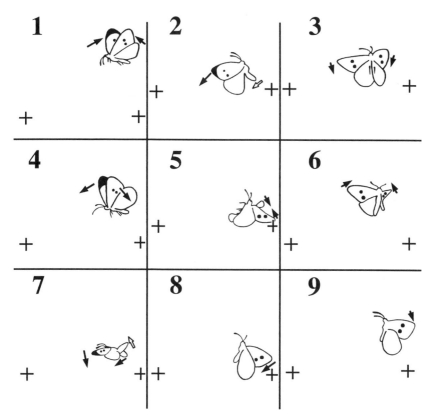

Figure 2.27. Sequential frames of a vertical maneuver of large white: lowering (1–3), upward turning (4, 5), and climbing (6–9). Black arrows, direction of wing movements; open arrows, direction of abdomen oscillations. The crosses correspond to the nodes of coordinate grid with distance between them 100 mm.

of mass. The butterfly prevents this by deflecting its abdomen upward in the lower part of each stroke, thus balancing against the torque.

Owing to its balancing function, the abdomen is used to alter flight direction from horizontal to ascending. To achieve this change, the butterfly needs only to hold the abdomen in line with the pterothorax at the bottom of the stroke. As a result, the body angle increases, and the butterfly starts climbing flight. Thus, the motions of the abdomen participate not only in horizontal turns (Camhi 1970; Kammer 1971; Zanker 1988a), but also trigger vertical maneuvers.

Coincidence of abdominal deflections with the end of downstroke and supination apparently shows that maximum lift occurs at the end of phase 2–

phase 3. One can therefore conclude that during horizontal flight, most lift is created in downstroke and supination, based on the data of wingbeat kinematics, patterns of vorticity, and in-stroke movements of the pterothorax and abdomen. In general, this is consistent with Nachtigall's (1966) opinion on the distribution of forces within the stroke cycle. Small insects, namely *Drosophila*, demonstrate wing kinematics different from large and medium-sized insects, even from a larger blowfly *Phormia regina* (Zanker 1990a). This is a probable basis for the difference of in-stroke force distribution found in *Drosophila* and is discussed subsequently.

Because the described mechanisms of force generation are inherently different, depending on situation, the relative contribution from each of them can be either increased or diminished. For example, in medium-sized and large insects, the circulatory and wake effects associated with wing translation are possibly predominant. By contrast, as the absolute dimensions of a flying organism decrease, the negative influence of viscosity on circulatory flow and vortex formation increases. In such a situation, rotary mechanisms may be advantageous because they are not obliged to vortex generation. Probably owing to the aerodynamic effects of wing rotation, the small fruit fly *Drosophila* exhibits two peaks of lift timed to pronation (squeeze-and-peel) and quick supination during each stroke (Zanker 1990b), although contradictory data exist; Dickinson and Götz (1996) recently observed the two lift maxima in *Drosophila* at the end of downstroke and in mid-upstroke. Presently it is difficult to explain these contradictions, but in any case the upstroke lift maximum—which could not be anticipated based on the data of wing kinematics and vortex dynamics—was reported for the first time. The authors provide an interpretation based on an extreme form of Wagner's effect (for more explanations, see Dickinson and Götz [1996]).

The exploitation of rotary mechanisms principally by small insects presumably has morphological consequences, namely, the origination of morphologically two-winged flight systems, because the use of rotary motion demands higher mobility of the wings, prohibiting coupling of hind- and forewings and causing reduction of the latter. Hence, morphologically two-winged insects are normally relatively small. Examples include mymarid wasps, scale insects, mayflies, and the order Muscida in general.

2.10. The Problem of Flight Power Regulation

While flying, insects are able to regulate power output and change lift and flight velocity within wide limits. Information on their methods of doing this is surprisingly contradictory.

According to published data, the nature of the relationship between flight speed and stroke frequency is quite ambiguous. The direct dependence of speed on frequency has been reported for tethered (Gewecke and Niehaus 1981) and free-flying (Sotavalta 1947; Betts and Wootton 1988) lepidopterans, tethered (Götz 1968; Gewecke 1974; Spüler and Heide 1978; Nachtigall and Roth 1983) and free (Sotavalta 1947) dipterous flies, and tethered orthopterans (Weis-Fogh 1964; Gewecke 1974, 1975, 1977; May et al. 1988). Other researchers have shown the absence of such interrelationships in tethered flies (Vogel 1967a; Cullis and Hargrove 1972) and free-flying butterflies (Grodnitsky 1993), and in dragonflies and damselflies (Wakeling and Ellington, 1997b). Reports on a correlation between speed and frequency are sometimes accompanied by data supporting different statements; for example, dependence is represented by a scattered cloud of dots (Baker et al. 1981). Contrary opinions have been suggested even for the same species, for example, housefly *Musca domestica* (cf. Cullis and Hargrove [1972] and Spüler and Heide [1978]).

A similar situation exists in bird flight research. For example, Tsvelykh (1986, 1988), Komarov and Mordvinov (1989), and Mordvinov (1992) argue that flight speed increases with the increase of flapping frequency. On the other hand, there is a series of articles showing that flight speed varies noticeably at a relatively constant frequency (Greenewalt 1960; Tucker 1966; Schnell 1974; Butler and Woakes, 1980; Tobalske and Dial 1996), and the opposite, that flight speed does not change under more frequent flaps (Pennycuick et al. 1989). In nightingale and teal flying in a wind tunnel, the frequency of wingbeats slightly but significantly increased over a particular minimum value at both an increase and a decrease of wind velocity (Pennycuick et al. 1996). Evidence also exists that separate flaps of the wings of the Andean condor sometimes generate only thrust and sometimes only lift (McGahan 1973). In bats both the presence (Carpenter 1985) and absence (Aldridge 1986) of this relationship have been reported.

The information on the relationship between flight speed and stroke amplitude (or stroke angle) is also contradictory. Positive correlation has been described for free-flying dragonflies (Alexander 1984; Rüppell 1989), butterflies (Grodnitsky 1992a, 1993), dipterous flies (Ennos 1989a), tethered beetles (Schneider and Krämer 1974), and fruit flies *Drosophila* (Vogel 1967a; Zanker 1988b). In tethered migratory locusts both positive (Gewecke 1975, 1977) and negative (Gewecke 1974) correlations have been observed. Inverse function has been found in free-flying (Betts and Wootton 1988) and tethered (Gewecke and Niehaus 1981) butterflies and in the tethered blowfly *Calliphora erythrocephala* (Gewecke 1974). For free-flying bumblebees *Bombus terrestris* (Dudley and Ellington 1990a) and bats (Aldridge 1986), the absence of correlation between flight speed and stroke angle has been reported. Wakeling and Ellington (1997b) found only a slight positive correlation between flight speed and stroke amplitude in the damselfly and dragonfly.

Direct correlation between flight speed and the downstroke-to-upstroke duration ratio has been reported several times, based on the results of studies of free-flying butterflies (Betts and Wootton 1988; Grodnitsky 1992a) and the coot (Komarov and Mordvinov 1989). At the same time, this correlation was found to be negative in free-flying hemipterans (Betts 1986c), the pigeon, and the magpie (Tobalske and Dial 1996). In free-flying bumblebees, neither a positive nor a negative relationship was observed (Dudley and Ellington 1990a).

Horizontal flight speed, as a rule, is negatively correlated to the angle of body inclination to the horizon. This is corroborated by numerous observations and experiments carried out on lepidopterans (Gewecke and Niehaus 1981; Dudley 1990; Grodnitsky 1992a, 1993), dipterous flies (David 1978; Götz and Wandel 1984; Zanker 1988b), bees (Esch et al. 1975), bumblebees (Dudley and Ellington 1990a), bugs (Betts 1986c), locusts (Weis-Fogh 1956a), and some birds, for example, the coot (Komarov and Mordvinov 1989). Nevertheless, some authors report an absence of interrelationships between flight speed and body angle in the tethered locust (Baker et al. 1981; Zarnack and Wortmann 1989).

In the majority of insect species, body angle is intimately related with the angle between stroke plane and the horizon. Thus, dependence of flight velocity on stroke plane angle must be of the same nature as the correlation between flight speed and body angle. Indeed, flight speed directly changes with the stroke plane angle in free-flying dragonflies (Rüppell 1989), dipterous flies (Ennos 1989a), bumblebees (Dudley and Ellington 1990a), bugs (Betts 1986c), and bats (Aldridge 1986). Information on the Lepidoptera is somewhat equivocal. For example, Betts and Wootton (1988), in the same article, noted a negative correlation for *Papilio rumanzovia* and *Graphium sarpedon* and a positive function for *Troides radamantus*. Studies of the natural flight of migratory locusts showed no distinct correlation (Baker et al. 1981).

Data on the relations between generated lift and different stroke parameters are no less contradictory. A positive correlation between lift and wingbeat frequency has been observed in free-flying dragonflies (Rüppell 1989), falcon (Videler et al. 1988), and the Harris' hawk (Pennycuick et al. 1989). In the tethered condition, it has been found in the small tortoiseshell butterfly (Gewecke and Niehaus 1981), fruit fly *Drosophila melanogaster* (Götz 1968) and blowfly *Calliphora vicina* (Nachtigall and Roth 1983), honeybee (Esch et al. 1975), bug *Triatoma infestans* (Ward and Baker 1982), and migratory (Gewecke 1975, 1977) and desert (Weis-Fogh 1964) locusts. The same correlation has not been found in free wasps (Sotavalta 1947), euglossine bees (Dudley 1995), and hummingbirds (Wells 1993; Chai and Dudley 1995, 1996; Chai et al. 1996) nor in tethered pomace flies *D. virilis* (Vogel 1967a). In some bugs it appears to be negative (Betts 1986c).

Lift is directly related to stroke amplitude in free-flying dragonflies (Alexander 1984), butterflies (Grodnitsky 1992a), orchid bees (Dudley 1995), falcon (Videler et al. 1988), and hummingbirds (Wells 1993; Chai and Dudley 1995, 1996; Chai et al. 1996). The same pattern characterizes the tethered flight of *Drosophila* spp. (Vogel 1967; Zanker 1988b), small tortoiseshell (Gewecke and Niehaus 1981), the bug *T. infestans*, and the migratory locust (Gewecke 1975). The opposite situation was noted in free-flying bugs of different species (Betts 1986c) and the tethered bluebottle fly (Nachtigall and Roth 1983). Lift and stroke angle are not correlated in free-flying migratory (Baker and Cooter 1979a) and tethered desert locusts (Wortmann and Zarnack 1993).

Dependence of lift on the body angle is somewhat less ambiguous. It was shown to be direct in free-flying butterflies (Grodnitsky 1992a), *Drosophila hydei* (David 1978), and bugs (Betts 1986c), and in tethered small tortoiseshell (Gewecke and Niehaus 1981), dipterous flies *D. melanogaster* (Götz and Wandel 1984; Zanker 1988b), *M. domestica* (Götz and Wandel 1984), and honeybee (Esch et al. 1975), and in desert locust (Wortmann and Zarnack 1987). This correlation was not observed in *T. infestans* (Ward and Baker 1982), and among tethered flying desert locusts, some specimens demonstrated a distinct covariation of lift with the body angle whereas others did not (Zarnack and Wortmann 1989).

Lift decreases with the growth of stroke plane angle in free-flying dragonfly *Megaloprepus coerulatus* (Rüppell 1989), in several bug species (Betts 1986c), and in a wasp (Kammer 1985). Nevertheless, this was not revealed during the study of *T. infestans* (Ward and Baker 1982), free-flying bees (Dudley 1995), and migratory locusts (Baker and Cooter 1979a).

Positive correlation between the angle of attack and lift was shown for tethered desert locusts (Wortmann and Zarnack 1987) but was not found in tethered migratory locusts (Baker and Cooter 1979a).

Angular velocity of wing movement is also correlated with generated lift in an equivocal manner. The correlation is positive in the dragonfly *M. coerulatus* (Rüppell 1989) and negative in bugs (Betts 1986c).

Increase of the downstroke-to-upstroke duration ratio makes a positive contribution to total lift in free-flying butterflies (Grodnitsky 1992a), bugs (Betts 1986c), and *Falco tinnunculus* (Videler et al. 1988). Nevertheless, it does not influence the lift value in migratory locusts (Baker and Cooter 1979a).

In general, there hardly exists a parameter of flapping flight that could be unequivocally correlated to the magnitude of lift and thrust generated by the wings during stroke cycle. Only wingbeat amplitude, in most cases, is directly related to lift. Contradictions in this field are so numerous and ubiquitous that they can be related to neither differences in experimental and observation procedures nor to the inaccuracy of researchers.

These disagreements are of a comprehensive character. Therefore, they are likely to be caused by a single reason, namely that numerous kinematic parameters participate in flight power regulation in insects and other flying animals. Besides the parameters previously discussed, several more characteristics should be mentioned on which sufficient comparative data have not been reported. These additional parameters include phase shift between the strokes of the fore- and hindwings (Alexander 1984; Somps and Luttges 1985; Rüppell 1989), angle of attack (Vogel 1967a; Dudley and Ellington 1990a), parameters of the wing trajectory (Zanker 1988b), wing deformation while flying (Ennos 1989a), and ratio of the durations and angular amplitudes of pronation and supination (Ennos 1989a; Dudley and Ellington 1990a). Flight power output can be regulated by changing the combination of particular values of these parameters, whereas single parameters can have no direct dependence on the power output.

The complexity of functional relationships in the system of flight power regulation suggests that each insect species possesses several stable kinematic patterns. Each pattern is characterized by a definite combination of the parameters of wing and body motion and probably corresponds to the least energy loss at any given flight speed. The patterns can also differ by the shape of the vortex wake formed behind a flying specimen (Rayner 1986). As one flight mode is being changed into another (e.g., slow to fast horizontal flight), separate parameters can be changing disproportionally to the flight speed change. Still, the function that describes the changes of each parameter can be specific in relation to wing morphology and shape, construction of the pterothorax, and other morphologic features. In all likelihood, this may explain the current inability to derive a general conclusion on the mechanisms of flight power regulation by considering the flight of representatives of different volant animal groups.

An example, corroborating the statement that simple linear correlations of kinematic parameters in insect flight power regulation systems should not be anticipated, is given in an excellent recent article by Lehmann and Dickinson (1997), who studied the correlation of stroke frequency and amplitude with the power generated in tethered flight of *D. melanogaster* and managed to demonstrate that minimal force is produced at very different combinations of amplitude and frequency. As the force increases, the combinations become progressively less scattered. Finally, the maximum force production corresponds to rather narrow intervals of both frequency and amplitude. The entire cloud of dots, marking the amplitude/frequency combinations, has the shape of a spiral half-coil with a sharp "large power" end and broad "small power" end. As a result, although stroke amplitude grows monotonically with power increase, flight force is positively correlated with wingbeat frequency at low power output and negatively correlated at high output, but, at the same time, the product of amplitude and frequency increases linearly with the force increase. This scheme can vary in different insects. Chai and coworkers (1997) recently reported very similar though less detailed data for free-flying hummingbirds.

2.11. The Problem of Vortex Wake Geometry

The first successful attempt to visualize the flow around a tethered flying insect was that of Demoll (1918, cited in Rohdendorf 1949), who described a general picture of the flow around a hawkmoth, averaged over a stroke cycle. More recently, use of thermoanemometers showed that during the course of a stroke cycle, the field of velocities around flying insects is not constant but undergoes regular pulsations, consistent with the frequency of strokes (Wood 1970; Bennett 1976). Current development of special methods provides an opportunity to study the nature of these pulsations and to describe the aerodynamic processes that follow the motion of the wings. These data allowed quantitative tracing of the development of the velocity field throughout the wing working cycle and acquisition of information about some finer points of interactions between flapping wings and the air.

The aerodynamic wake of a flying insect can be divided into near and far regions. The near wake contains vortex structures formed during the current stroke that interact with the body and wings. The far wake includes vortices formed during past strokes, which departed the flying specimen and do not directly interact with it.

2.11.1. Near Vortex Wake

Generally, at the beginning of a stroke, each wing (or wing couple) generates a starting vortex tube closed by one end to the wing tip and by the other to the trailing edge. As the wings move downward, the single tubes enlarge and soon merge, forming a single U-shaped bubble closed to the wing tips. After supination, when the trailing edges pause to begin the upstroke, the stopping vortex tubes are shed from the wings. These tubes are connected to the upper U-shaped bubble by their distal ends. The proximal parts of the stopping tubes are closed to the trailing edges of flapping wings. The stopping tubes merge with each other under the insect body during the upstroke. As a result, a complete self-closed vortex ring is formed. Throughout the remaining part of the stroke cycle, the ring remains connected to the wings, and leaves them only under the effect of fanlike throw-off during pronation. Thus, during each stroke the wings of a flying insect normally generate a single vortex ring. In the process of its formation, the ring can be far from toroidal in shape. In addition, different ring parts have different transverse diameters and, in all likelihood, different rotation speeds. Hence, different parts of the ring are characterized by different impulse values that stem from the complex character of the wing motion and the short duration of the stroke cycle. During downstroke, vorticity is stronger owing to larger angles of attack. In the course of upstroke, this angle is considerably diminished, with a consequent reduction in the intensity of vortex formation (see also Ellington et al. [1996] and Willmott et al. [1997]). Next, at the uppermost trajectory point, the wings approach each other closely or mutually adjoin. At the lowest point, this normally is not achieved. Consequently, starting tubes merge comparatively early, whereas the stopping bubbles connect relatively late. Hence, the right and left starting vortices enhance each other, whereas the stopping tubes are partially deprived of such opportunity and are decelerated to a larger degree by the air viscosity that slows down rotation of the tubes. The difference in strength between the starting and stopping tubes results in a pronounced uneven distribution of vorticity within the entire ring at the first stages of its existence.

Visualization results show that the ring becomes more uniform and acquires a regular shape after losing its connection with the wings. In addition, the diameter of its circular axis essentially decreases. Compression of rings after the end of the stroke cycle was theoretically anticipated (Rayner 1979a) and previously reported (Brodsky 1988, 1991).

2.11.2. Far Vortex Wake

Presently three main models of the far wake of flying animals exist (Alexander R. McN. 1986). These models describe flight at different values of advance ratio (J), which represents the ratio of flight velocity to mean flapping velocity (Ellington 1984). The advance ratio also determines the wing path shape. At high advance ratios, the wing tip trajectory becomes elongated. Therefore, at larger J, the wing can operate with positive angles of attack throughout the entire stroke cycle (fig. 2.28A), thereby preventing the formation of stopping tubes at the bottom of the trajectory. If larger J values are accompanied by low stroke amplitude, when the wings of different sides do not meet each other above the body, connection of the starting tubes does not take place, and at any moment of the stroke, the vortex wake is represented by two vortex cords or filaments coming from the wing tips. A wake of such type is formed during the fast straight flight of birds (Rayner 1986; Spedding 1987), bats (Rayner et al. 1986) and insects (Willmott et al. 1997). It differs from a typical aircraft wake (see fig. 2.5B) by the spatial curvature of the tubes. Flapping flight with this kind of wake geometry is considered easier to analyze and model, because nonstationary effects such as periodic interactions of the wings with evoked vorticity can be neglected, in contrast to flight with the following two types of wake geometry (Robinson and Luttges 1983; Ohmi et al. 1990; Send 1992).

Wings have negative angle of attack and generate negative lift throughout upstroke at small J, owing to peculiarities of their morphology that limit the degree of supination. Hence, as the sign of the angle of attack changes during supination, the stopping tubes are shed from the trailing edges, and at the same time function as starting vortices for the upstroke. Merging of the starting and stopping tubes engenders the ring structure of the wake. Since

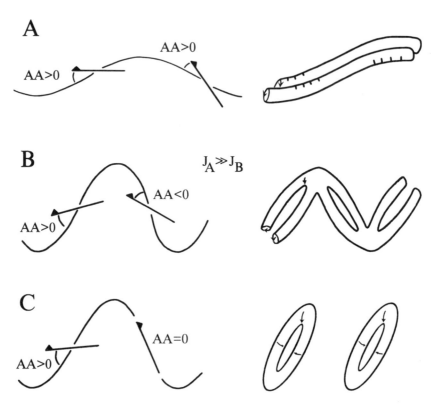

Figure 2.28. The shape of the far vortex wake, depending on the sign of the angle of attack (AA). J = advance ratio. (A) Two vortex tubes; (B) connected vortex rings; (C) separate rings.

the wings are aerodynamically efficient throughout the entire stroke, tip vortices constantly come from their surface, ensuring continuity of the wake. Accordingly, the wake looks like a chain of rings inclined to each other and linked by their upper and lower parts (fig. 2.29B). This wake structure is derived from the hovering and slow forward-flight kinematics of insect wings (Brodsky and Ivanov 1983a, 1984; Brodsky 1984, 1986b, 1988, 1990, 1991; Brodsky and Grodnitsky 1985; Rayner 1986; Ivanov 1989, 1990).

In contrast to insects, bird wings can perform upstroke in a half-folded condition that is aerodynamically inefficient. This results in discontinuity of the wake, which, in this case, consists of separate vortex rings (fig. 2.29C) (Kokshaysky 1979, 1982; Kokshaysky and Petrovsky 1979a, 1979b; Spedding et al. 1984; Spedding 1986). The same wake shape is possible if the wings in

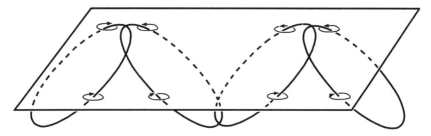

Figure 2.29. The model of connected vortex rings and the flow picture that must occur at cross-sectioning of the wake by a horizontal light plane situated above the insect.

the second half-cycle possess zero angle of attack, so that forces are not generated and vortices are not formed.

The spatial structure of the aerodynamic wake proposed in the section 2.11.1 for hovering and slow forward flight differs from the model of linked rings, which is more generally accepted in the descriptions of insect flight. According to the material presented, the wake consists of independent rings, each of which is formed at the hindwings' supination and completes its formation by the end of the stroke. Similar data were obtained for takeoff of the large white (Ellington 1980) and the tethered flight of fruit flies (Dickinson and Götz 1996).

The conflicting interpretations of experimental data require additional argumentation for the suggested wake shape. I propose that although wings are aerodynamically active throughout the upstroke, and connected rings should therefore be anticipated, separate rings appear as the result of the wingbeat cycle asymmetry. The model of separate rings is chosen because the cross section of a series of linked rings by a horizontal plane, situated above or below the insect, must inevitably present a pattern containing groups of four vortices brought together (fig. 2.29). This type of flow field has never been reported by any of the researchers who defend the model of linked rings.

Why are separate rings formed instead of connected ones? First, the left and right wings normally do not meet in the bottom of their trajectory. Consequently, rings of sequential half-strokes have a large common part that equals the entire lower U-shaped bubble and represents up to one-third of the ring's circular axis. Thus, initially the rings of downstroke and upstroke are partially united. This favors coalescence of the two rings in the course of the cur-

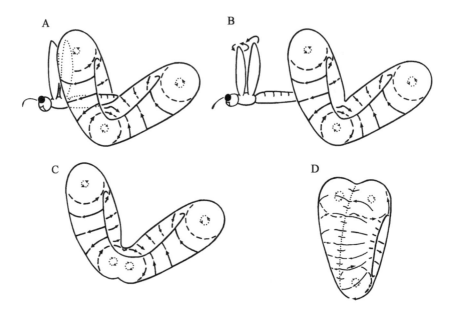

Figure 2.30. Different wake types anticipated for free flight. Vortices are shown that are generated within one wingbeat cycle. (*A*) V-likely connected rings at the end of upstroke; (*B*) same, after the beginning of the next stroke; (*C*) separating rings; (*D*) coalescing rings.

rent stroke (as in fig. 2.30*D*). In other words, vortex filaments that come from the wing tips during upstroke are accumulated by the vortex ring generated in the course of downstroke. This is supported by the results of a physical experiment (Maxworthy 1972), showing that initially even completely separate rings have a tendency to merge. Contrary to theoretical predictions (Fabrikant 1964; Rayner 1979a), vortex rings do not assemble in groups and do not pass into one another. As a result, a single ring results from each stroke cycle.

Second, in insects with narrow wing hinges, the dorsal vortex (and hence the entire upper U-shaped bubble) does not change its position over the body throughout the working cycle. Therefore, two rings that could be formed during the first and second half-cycles are combined in space. Generation of two rings per stroke is prohibited by spatial limitations: each ring has a relatively large cross-section diameter. It is consequently impossible for two rings to simultaneously occupy the small space that the insect flies through during a single stroke.

Separate rings provide the least controversial reconstruction of the wake spatial structure based on the data from visualization experiments. However,

Figure 2.31. Left wing tip path of the blowfly *Calliphora erythrocephala* in an independent coordinate system. (After Nachtigall 1985.)

in fast forward flight during downstroke, the dorsal vortex moves away from the body to a distance sufficient for the formation of a separate ring of the second half-cycle. As a result, each stroke generates two V-like connected rings (fig. 2.30*A, B*). An analogous conclusion is derived from the results of smoke visualization of flow around the peacock butterfly (Brodsky 1990, 1991) and the tobacco hawkmoth (Ellington et al. 1996; Willmott et al. 1997): At high flight velocity the rings part afterward (fig. 2.30*C*) (see Willmott et al. 1997); at low forward speed they merge (fig. 2.30*D*). Meanwhile, formation of the wakes of such shape in free flight is not inevitable, because the wing path relative to the air is such that the wing completes downstroke and begins upstroke practically in the same space (fig. 2.31), because supination, compared with other stroke phases, is the period when the body travels the least horizontal distance (fig. 2.26*B*). Therefore, for the entire lower part of the stroke cycle, the wings move inside the zone of their own induced vortex flow. Thus, prerequisites are created not for partitioning, but for coalescence of the rings of downstroke and upstroke.

In hovering and in slow flight, vortices of different strokes are not connected with each other. Segregation of the rings arises because the dorsal vortices of sequential strokes are separated in time and space: Having closed to itself, a vortex tube forms a ring, loses its connection with the wings, and is carried away by the flow generated by fanlike air throw-off during pronation. Conversely, new tubes begin to form between the upper sides of the wings only after the greater part of pronation, when the insect has already flown a significant distance, because pronation is the phase of the longest horizontal travel (fig. 2.26*B*).

Finally, any cycle of vortex formation has the same result, despite kinematic differences. Owing to various reasons, the wake always appears as a series of separate rings, as was presupposed by quantifiable models (Rayner

1979a; Ellington 1980, 1984). The most essential difference separating the proposed viewpoint from the existing vortex theories of insect flight is that the rings generated during natural flight cannot be considered as thin, as is postulated in the cited articles. Measurement of the rings in free flight showed that the relationship between the diameter of the ring cross section and the diameter of the ring circular axis after the stroke cycle is finished varies from 0.18 to 0.38, and decreases when the ring is leaving the insect and its rotation is slowed down under the effect of air viscosity. Judging from visualization data, the wake lasts approximately 0.1 s, and then the vortex rings disappear.

2.11.3. Double Vortex Chains

According to Brodsky's (1985, 1986b, 1988, 1991) assumption, an increase in stroke frequency up to a critical value of about 70 Hz leads to the formation of a separate vortex chain behind each wing. This concept contains internal contradictions: the presence of rings at each body side must induce an airflow over the tethered insect directed from the abdomen tip to the head. However, such a reversal has never been found (Grodnitsky and Morozov 1993). The results obtained indicate that the flow around crane flies (see sec. 2.4) does not bear the striking difference suggested by Brodsky. Furthermore, even after a considerable increase in stroke frequency (up to 100–120 Hz in bugs and dipterous flies), the wings generate only one vortex ring per stroke.

It can be assumed that the flow structure was interpreted as double chains because flow pictures taken from above contain two vortices on each side of a flying insect. The conclusion of a single ring at each side of the body is quite evident, and Brodsky's opinion is therefore understandable. However, a more detailed analysis of the vortex wake shows that a typical dorsal vortex is present over the body of the crane fly. The double chains model does not include the dorsal vortex. Instead, flow reversal must occur. The data should be interpreted as though these vortices result from the cross-sectioning of two strongly curved stopping tubes by a light plane, so that the plane strikes each of the tubes twice (fig. 2.12G, H). Accordingly, the two tubes appear on photographs as four two-dimensional vortices.

Nevertheless, the idea of double chains appears to be fruitful. A wake of such shape is formed during some strokes of free-flying butterflies. The formation of a separate vortex wake behind each wing or wing couple is possible if the right and left starting tubes do not merge above the body at the beginning of the stroke cycle, that is, if they have small rotation velocity and are separated by a space large enough to prevent the formation of an upper U-shaped bubble. This can happen if the wings move with a relatively low speed, performing incomplete strokes, and do not approach each other closely in the upper part of their trajectory. In this case, stagnation of dust particles is seen in the space between the two vortex rings, exactly as the specific character of the discussed model requires (in tethered flight in still air the stagnation is expressed as a flow reversal). However, the reason for the formation of a separate wake behind each flapping wing/wing couple is not a strengthened stroke, as was suggested by Brodsky, but just the opposite, a weakened wingbeat modification with a reduced trajectory top (Grodnitsky and Dudley 1996). Dickinson (1996) suggests identical reasons for the formation of a double wake.

2.12. Conclusions

Examination of the phenomenon of insect flapping flight shows that one working cycle of the wings consists in a succession of different aerodynamic processes separated in space and time. Circulation and interaction with the vortex wake take place during the wings' translation, whereas fling and fanlike throw-off follow rotary movements (pronation and supination). The relative contribution from each mechanism to the resulting aerodynamic force is unknown, but presumably can vary in insects of different size. Thus, in medium-sized and large insects, lift maximum coincides with the end of downstroke and supination, whereas in small forms such as *Drosophila*, within-stroke lift distribution is different.

In all likelihood, insects are able to gain energy from their own vortex wake, controlling the interaction of wings and generated vortices. The near vortex wake may therefore be represented as an accumulator that periodically takes and returns energy during the wingbeat cycle. This statement complements common insight regarding cyclic accumulation of energy during flapping flight. Thus, the concepts of weight (Borin and Kokshaysky 1982) and elastic (Boettiger and Furshpan 1952; Weis-Fogh 1959, 1960, 1965; Pringle 1963; Wilson 1965; Weis-Fogh and Andersen 1970; Pfau 1973, 1977, 1985, 1987; Thomson and Thompson 1977; Wisser and Nachtigall 1983; Ellington 1984; Miyan and Ewing 1985, 1988; Bennet-Clark 1986; Ennos 1987) energy accumulators are augmented by the third one, the wake energy accumulator. The process of evolutionary formation of the cyclic system of energy accumulation can provide a field for special examination. There is presently insufficient comparative material on this problem, but strong controversies nonetheless exist. Thus, elaboration of a general view on periodic energy storage in insect flying systems is currently impossible. However, the improvement of interactions between the wake and wings defined the main trends in the evolution of the flapping-flight function in insects. Chapter 3 considers the evolution of flight in the insect orders.

3 | Evolution of Flight in the Insect Orders

Flight is not the only function of wings and their derivatives. Apart from their role in locomotion, wings serve as defense organs, are used to regulate body temperature, and often carry diverse signal color patterns (Rohdendorf 1949). Different wing functions may be incompatible with one another. For example, flapping flight is more economical with light wings, whereas defense demands wings of greater mass. The resulting adaptive complex reflects a set of compromises in which each function is fulfilled only to the extent that it does not prevent the execution of another (Rasnitsyn 1987).

The evolution of flight in each order was influenced by specific preexisting (or evolving) features of construction and/or mode of life. The presence of a typical functional and morphologic environment also conditioned the typical course of flight function development. Therefore, flight was improved in a peculiar manner in each order.

My own experience is based on studies of endopterygote insects (cohors Scarabaeiformes), a group that embraces the main functional and structural diversity of flight systems. Conclusions on the wing functional morphology of other taxa were derived from published data on dragonflies (Chadwick 1940; Zalessky 1951; Sotavalta 1954; Neville 1960; Hisada et al. 1965; Ryazanova 1966, 1968; Norberg 1972a, 1975; Pfau 1986; Pond 1973; Rudolph 1976a, 1976b; May 1981, 1991; Pfau 1982, 1986; Alexander D. E. 1984, 1986; Bocharova-Messner and Dmitriev 1984; Azuma et al. 1985; Rüppell 1985, 1989; Somps and Luttges 1985; Newman and Wootton 1986; Azuma and Watanabe 1988; Wootton 1991), mayflies (Edmunds and Traver, 1954; Brodsky 1971, 1974, 1975, 1981a), locusts, crickets, and grasshoppers (Jensen 1956; Weis-Fogh 1956a, 1956b; Weis-Fogh and Jensen 1956; Gettrup and Wilson 1964; Dugard 1967; Waloff 1972; Zarnack 1972, 1975, 1983; Cooter 1973; Svidersky 1973; Cooter and Baker 1977; Gewecke 1977, 1983; Pfau 1977, 1978; Baker 1979; Baker and Cooter 1979a, 1979b; Baker et al. 1981; Kutsch and Gewecke 1981; Nachtigall 1981a, 1981b; Pfau and Nachtigall 1981; Thüring 1986; Schwenne and Zarnack 1987; Wortmann and Zarnack 1987, 1993; Banerjee 1988; Brackenbury 1990, 1991a, 1992; Schwenne 1990; Wilkin 1990), stoneflies (Brodsky 1979a, 1981b, 1982, 1986a, 1988), and rhynchote insects (Barber and Pringle 1965; Hewson 1969; Govind and Burton 1970; Pouchkova 1971; Ohkubo 1973; Gringorten and Friend 1979; Baker et al. 1980; Ward and Baker 1982; Betts 1986a, 1986b, 1986c; Wootton and Betts 1986; Byrne et al. 1988; Wootton 1988, 1996; Brackenbury 1992).

3.1. The Origin of Flight

The problem of the origin of insect flight is in fact the problem of the origin of insects themselves, because the winged insects Scarabaeona encompass the absolute majority of known recent and extinct species. Any theory that satisfactorily describes the origin of insect flight must provide answers to two basic questions: (1) What were the primary structures from which wings originated?, and (2) What factor caused the initial increase in the wing precursor? Because the wings must be of sufficient size for active flight, initially small vestiges could not perform as aerial propelling agents. Thus, the initial function of winglets was definitely not flapping flight, and flight origin instead constituted a typical case of function change (see Dohrn 1875). A search for the solution to the two essential questions has engendered numerous assumptions about the structure and functions of proto-wings (most recently reviewed by Kingsolver and Koehl [1994] and Dudley [1999]).

Some authors believe that wings originated from articulated organs such as gills (Wigglesworth 1963, 1973, 1976; Mamaev 1975; Leech and Cady 1994; Marden and Kramer 1994, 1995; Kramer and Marden 1997), parts of gills (Birket-Smith 1984) or similar structures (Carle 1982), respiratory appendages of legs (Kukalova-Peck 1978, 1983, 1987, 1991, 1992; Trueman 1990; Averof and Cohen 1997), prototergalia of body segments (Kliuge 1989), or spiracle valves (Bocharova-Messner 1968; Brodsky 1979b). Others consider wings to be derivatives of nonarticulated organs such as lateral (paranotal) (Martynov 1938; Forbes 1943; Shwanvich 1946, Zalessky 1949, 1953; Hinton 1963; Flower 1964; Rasnitsyn 1976a, 1981; Wootton 1976) or posterolateral (Becker 1966) projections of the dorsal part of thoracic segments.

The evolutionary enlargement of wing vestiges could be caused by one of three reasons related to the mode of life of pterygote insect ancestors.

1. The first winged insects were small; the appendages favored passive transfer by wind (Wigglesworth 1963, 1976; Norberg 1972b). Some authors hypothesize that the original insects lived in arid and semiarid conditions in which heat convection caused strong elevating air currents to flow at up to several meters per second (Rainey 1965).

2. The appendages helped to run, performing fast flaps (Shwanvich 1946; Kukalova-Peck 1987) or balancing movements (Brodsky 1979b), or they served for skimming on the water surface, functioning first passively without strokes, as sails under wind (Marden and Kramer 1995), and then becoming active flapping organs of aerodynamic propulsion (Marden and Kramer 1994).

3. The first Pterygota often jumped from elevations and then flew by gliding; the appendages potentially extended the flight range (Snodgrass 1935; Martynov 1938; Forbes 1943; Hamilton 1971; Rasnitsyn 1976a, 1980; Wootton 1976; Wootton and Ellington 1991).

In comparing these viewpoints, three important facts should be mentioned. First, there are no paleontologic data indicating that the primitive insects that evolved into pterygotes were small. Rather, they were medium-sized or even large (Rohdendorf and Rasnitsyn 1980). Second, organisms with six legs do not need balancing when they run (Emelianov and Falkovich 1989). Moreover, balancing movements, as opposed to flapping, are always directed contrary left-to-right. Third, the life mode of many generalized recent insects, including many apterygotes, mayflies, dragonflies, stoneflies, alderflies, caddisflies, primitive beetles, and dipterous flies, is closely connected with water (Tikhomirova 1991). The aquatic habits of these extant insects is one of the main reasons for the assumption that primary insects lived in water. Nevertheless, the ancestors of pterygote insects evidently were neither aquatic nor semiaquatic. They had descended from an aquatic myriapod-like ancestor, but underwent further evolution in terrestrial habitats (Rohdendorf and Rasnitsyn 1980). This view is supported by the absence of any paleontologic data indicating an aquatic life mode of the first insects. Even the primitive Libellulida are thought to have terrestrial nymphs, since no one fossil nymph is known from the Carboniferous lakes, whereas lots of imago have been found. In general, insect nymphs with definite adaptations to life in water appeared only in the Permian era (Rohdendorf and Rasnit-syn 1980). However, the crucial fact is that all recent and extinct aquatic insects breathe with tracheal gills, which are a secondary adaptation to life in water, because the primary tracheal system was constructed for air breathing (Ghiliarov 1949, 1957; Becker 1952; Tikhomirova 1991). This evidence strongly contradicts all hypotheses in which the wing originates from gills and gill-like structures, a currently popular belief in the West.

The presence of stroke ability in the wing precursors is also disputable. It suggests that a powerful muscle motor could be generated before the origin of flight, because the generation of useful force by small wings requires high flapping frequency (see also Borin 1987). From theoretical considerations (Casey 1989) and experimental data (Feller and Nachtigall 1989), energy expenditures and total power output are noticeably higher during hovering than horizontal flight; in fact, energy expenditure is normally a U-shaped function of flight speed (Schmidt-Nielsen 1972; Pennycuick 1975; Rayner 1979b; Dudley 1992). Only bumblebees are exceptions to this pattern owing to specific features of their flight kinematics (Dudley and Ellington 1990a; Ellington et al. 1990).

The energetic value of takeoff is unknown but is apparently close to or exceeds that of hovering flight, which suggests that the first flight mode to appear in evolution was straight flight because it is the easiest form of aerial locomotion. Later, with the appearance of a more powerful wing motor, the majority of insects acquired the capacity to overcome their own inertia and to take off from the ground. In many groups of insects, the difficulty of take-off conditioned the evolution of auxiliary jump mechanisms, which launched their bodies into flight. Certainly flapping wings can have a selective value in downward jumps, participating in the control of fall direction (Birket-Smith 1984; Brodsky 1988), as in the German cockroach. However, this assumption still fails to answer the cause of proto-wing enlargement to an aerodynamically efficient size.

With respect to the problem of the wing morphologic precursor, there are no serious reasons to deny the paranotal nature of the wings (Emelianov and Falkovich 1983). This evolutionary scenario is especially attractive because flight could emerge before a complex muscle motor and differentiated pterothorax skeleton were formed. Logically the best historically and functionally corroborated concept of the origin of flight is the modification of the paranotal theory presented by Rasnitsyn (1976a, 1980, 1981) (fig. 3.1). According to this hypothesis, the ancestors of winged insects inhabited comparatively

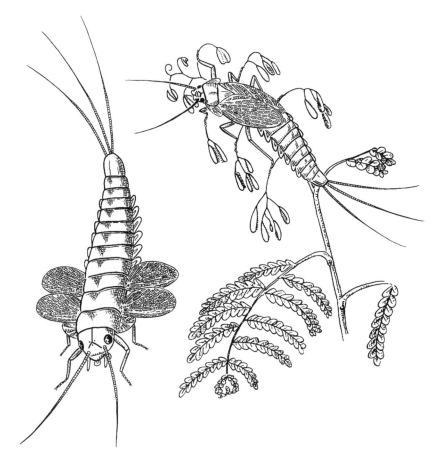

Figure 3.1. Hypothetical ancestors of winged insects. (From Rasnitsyn 1976.)

tall vegetation. In all probability, these organisms could already behaviorally regulate body temperature. Their paranotal lobes were theoretically adapted for the accumulation of solar heat (Douglas 1981; Kingsolver and Koehl 1985, 1989, 1994), which favored the initial increase of these lateral thorax projections. The ancestors of winged insects jumped from branch to branch while seeking food and avoiding attack by predators. In this situation, selection could favor further elongation of the paranotal lobes, which gradually acquired the function of bearing planes.

Plaiting (or folding up) of the primary wings supposedly originated simultaneously with their elongation. If it had not, long paranotal lobes would have considerably handicapped insect movement among tree branches (Shwanvich 1946). Among recent insects, similar behavior is exhibited by the North

American grasshopper *Pterophylla camellifolia* (Fabr.) (Gryllida: Pseudophylli-nae) (Alexander and Brown 1963). However, note that articulated paranotal lobes are present in some recent crustaceans and acari (Shwanvich 1946; Za-lessky 1949). Acquisition of wing-folding ability provided prerequisites for the development of their motility. Insects acquired the capacity to change angle of attack and, consequently, to extend flight or to perform turns in a de-sired direction. Strokes could originate next, first low-amplitude, serving to augment starting vortices, and then more energetic. Although the functional properties of bearing planes are aggravated after the appearance of strokes (Caple et al. 1983), this disadvantage could be compensated, for example, by increased maneuverability or other capacities not necessarily concerned with wing apparatus and flight. The origin of insect flight could also have been facilitated by a marked increase of atmospheric oxygen (Graham et al. 1995, 1997).

3.2. Initial Wing Kinematics

Rasnitsyn (1996) described the general procedure for tracing phylogenies as a set of presumptions. According to one of them, the least functionally efficient morphologic state of a given structure should be considered as the earliest and most primitive unless there is sufficiently strong evidence supporting a different viewpoint.

Which of the malfunctional kinematic patterns corresponds to the begin-ning of flapping-flight evolution? The functionally two-winged flight of tiger beetles and some swallowtail butterflies, with uncoupled fore- and hind-wings, and the counterstrokes of dragonflies and damselflies are not aero-dynamically advantageous (see sec. 3.5), but they are clearly secondary phe-nomena derived from wingbeats typical of these taxa.

According to the data presented in section 2.6, the first stroke cycle stage in anteromotoric, functionally four-winged in-phase flappers is accompanied by the appearance of a separate vortex tube behind each wing (fig. 2.20). Vor-tices of the fore- and hindwing are located close to one another. Since the direction of their rotation is the same, they reduce one another's velocity and actually heat the surrounding air. Afterward the tubes coalesce, and a U-shaped bubble is formed. However, starting deceleration must negatively influence further circulation growth and vortex wake development, decreas-ing the total aerodynamic force generated by the wings.

In this situation, the fore- and hindwings constitute competing generators of vortex flow. Therefore, this is a maladaptive, presumably primitive state. This assumption is further supported by the fact that not one taxon character- ized by anteromotoric in-phase kinematics (Panorpida, Perlida, Myrmeleon- tida, Corydalida, Raphidiida, Embiida, caddisflies Rhyacophilidae, primary moths Micropterigidae and Eriocraniidae) includes species that possess flight capacities normally considered to be advanced, that is, with long duration, high velocity, ability to carry load in flight, or pronounced maneuverability. Zorotypida, which have very archaic flight apparatus morphology (meso- thorax independent of metathorax, the former larger than the latter; fore- wings larger and better developed than the hindwings; a coupling mecha- nism is absent) seem to constitute a group in which flight reduction is in progress. Although it is believed that these insects are capable of active dis- persal and flight, given their life in rotten wood, no one has ever seen them flying (Choe 1992; personal communications of R. C. Nelson and G. T. Riegel). Remarkably, the orders of typically anteromotoric four-winged insects are among the least numerous and diversified orders of the class Insecta (Borror et al. 1992).

The wingbeat kinematics of lacewings (Ellington 1984), scorpion flies (fig. 3.2A), and stoneflies (fig. 3.2B) are quite similar. These insects are character- ized by comparatively low (from 15 to 50 Hz) stroke frequency (fig. 3.3) and wing loading (fig. 3.4), and by medium wing couple aspect ratio (fig. 3.5). Dur- ing the stroke cycle, the wings of these insects move in a plane almost per- pendicular to the longitudinal body axis. The forewings are not coupled with the hindwings and perform independent strokes. At the end of each stroke cycle, the wings meet above the body in a movement termed *clap* (Weis-Fogh 1973). Owing to the clap, the forewings are motionless at the very beginning of the stroke, tightly adjoining each other by their dorsal surfaces. After the stroke begins, the forewings pronate and open in a book-like manner (Grod- nitsky and Morozov 1994). This is the fling mechanism as initially described by Weis-Fogh (1973). Wing translation occurs only after fling finishes.

Strokes of the fore- and hindwing pairs are synchronized in a specific manner. The forewings constantly move ahead of the hindwings, but display a smaller amplitude. Hence, when deceleration and supination of the fore- wings begin in the lower part of the trajectory, the hindwings leave them behind, performing a more complete stroke (fig. 3.2). Therefore, functionally four-winged insects possess a specific aspect of stroke asymmetry absent in species with coupled wings: downstroke is made together (the wings, although not linked morphologically, behave as a single flapping surface),

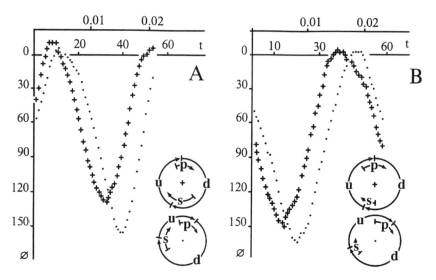

Figure 3.2. Wingbeat kinematics of functionally four-winged insects. (*A*) Scorpion fly *Panorpa communis* L.; (*B*) stonefly *Alloperla mediata* Navas. Horizontal: (t) time—upper scale, seconds; lower scale, frames. Vertical: (∅) wing positional angle (°). (+) forewings, (·) hindwings. Angular diagrams indicate interrelationships between different wing motions: (d) downstroke, (u) upstroke, (p) pronation, (s) supination. Upper diagram shows the stroke cycle of forewings; lower one, hindwings.

whereas upstroke is fulfilled separately. Strictly speaking, these insects are functionally four-winged only through the period of the second half-stroke. The trajectory of the hindwing pair is reduced above the body. In contrast to the forewings, the hindwings touch only slightly over the body, and then begin a new stroke cycle.

In neuropterans and scorpion flies, the wings remain completely flat throughout the first half-stroke. During supination, as a rule, only a smooth profile concavity can be seen. In species with comparatively broad wings (e.g., lacewings), bending along furrows is observed: for example, along the remigio-anal of all four wings in *C. dasyptera* (Grodnitsky and Morozov 1994) and along the remigial furrows of the forewings in *Chrysopa carnea* (see photo in Dalton [1975]; Wootton [1990]). These bends persist throughout the entire second half-stroke. In *Panorpa germanica,* the distal parts of both wing pairs bend ventrally at the lower trajectory point and remain in this condition during the initial part of the upstroke (Ennos and Wootton 1989).

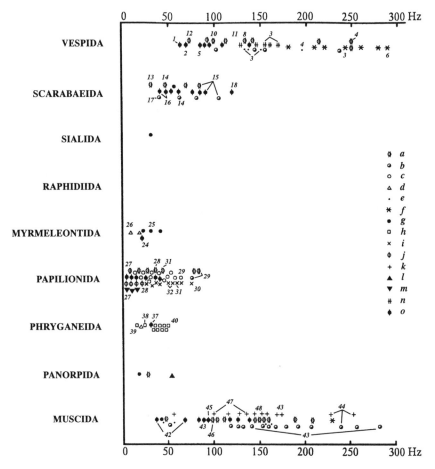

Figure 3.3. Stroke frequency (Hz) in different endopterygote insects. Data are from the following: a, Magnan (1934); b, Weis-Fogh (1973); c, Bartholomew and Casey (1978); d, Antonova et al. (1981); e, Ellington (1984); f, Unwin and Corbet (1984); g, Brodsky (1985); h, Ivanov (1985a); i, Kozlov et al. (1986); j, Betts and Wootton (1988); k, Ennos (1989a); l, Ennos and Wootton (1989); m, Dudley (1990); n, Dudley and Ellington (1990a); o, Grodnitsky and Morozov (1994). Subordinate taxa: Vespida (1–12), Scarabaeida (13–23), Myrmeleontida (24–26), Papilionida (27–36), Phryganeida (37–41), Muscida (42–51); 1, *Tenthredo;* 2, Sphecidae; 3,'*Bombus;* 4, *Apis;* 5, *Ancistro-cerus;* 6, *Trigona;* 7, Ichneumonidae; 8, *Xylocopa;* 9, *Arge;* 10, *Vespa;* 11,'*Ammophila;* 12, *Allantus;* 13, *Lucanus;* 14, *Melolontha;* 15, *Cetonia;* 16, Cerambycidae; 17, *Heliocopris;* 18, *Mordellistena;* 19, Coccinellidae; 20, *Trichius;* 21, different Scarabaeidae; 22, Cantharididae; 23, Curculionidae; 24, *Chrysopa,* 25, *Hemerobius;* 26, *Myrmeleon;* 27, broad-winged species from different lepidopteran superfamilies (Geometroidea, Saturnioi-dea, Papilionoidea); 28, Noctuidae; 29, Sphingidae; 30, Aegeriidae; 31, Zygaenidae; 32, Micropterigidae and Eriocraniidae; 33, Cossidae; 34, Pyraloidea; 35, Saturniidae; 36, Heliconiidae; 37, *Apatania;* 38, *Rhyacophila;* 39, *Semblis;* 40, *Hydroptila;* 41, *Anabolia;* 42, Tipulidae; 43, Syrphidae; 44, *Drosophila;* 45, Asilidae; 46, Tabanidae; 47, *Bibio;* 48, Calliphoridae; 49, Simuliidae; 50, Culicidae; 51, Muscidae.

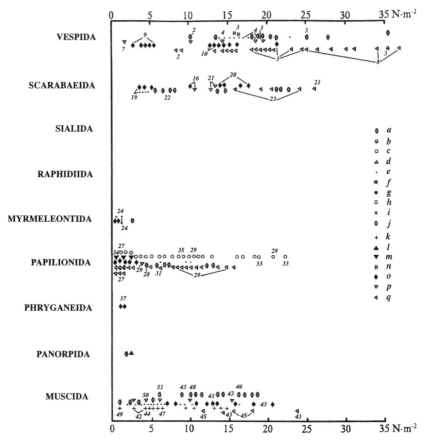

Figure 3.4. Wing loading (N/m²) in endopterygote insects. Data sources as in figure 3.3; also p, Sotavalta (1952) and q, Marden (1987). 1–51 as in figure 3.3.

3.3. The Origin of Major Kinematic Patterns

The effectiveness of primary flight kinematics could be enhanced by eliminating the interaction between the starting vortices of the fore- and hindwings. This could be achieved if a dominant wing pair generated more intensive vortices that were less susceptible to deceleration, or even completely suppressed vortex formation over the adjacent wing pair (Grodnitsky 1994, 1995a).

Two alternative solutions can be imagined: (1) generation of vortices by the effort of the forewings, or (2) of the hindwings. The first stage of such a process could consist in a decrease of vortex formation behind one wing pair and vorticity reinforcement over the other. To weaken a vortex, it is nec-

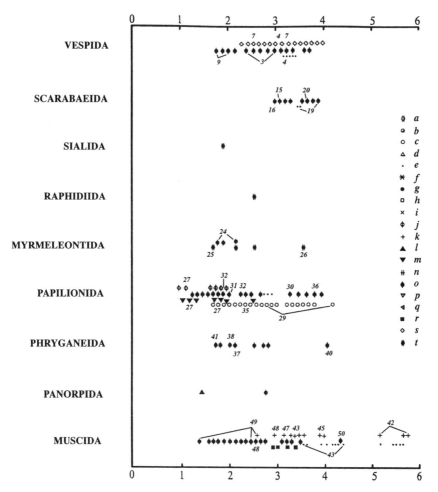

Figure 3.5. Wing/wing couple aspect ratio in endopterygote insects. Data sources as in figures 3.3 and 3.4; also r, Ennos (1988); s, Danforth (1989); and t, original calculation according to the drawings in Comstock (1918).

essary to decrease the wing size and/or to reduce its trajectory. Both these strategies were used during the course of insect flight evolution.

The first anteromotoric strategy can be realized by means of two methods. The first consists in augmenting the forewing strokes, weakening the hindwing strokes, and/or diminishing the size of the hindwings (fig. 3.6B). Fossil scorpion flies, whose hindwings were considerably decreased, typify this particular flight system (Rohdendorf and Rasnitsyn 1980; Shcherbakov et al. 1995). Scorpion flies with shortened hindwings are suggested to have

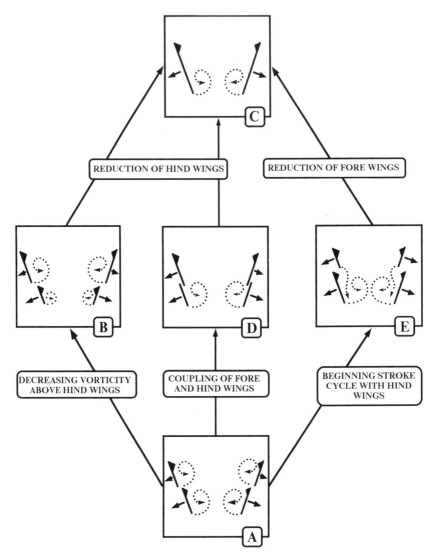

Figure 3.6. Evolutionary adaptive changes of wing kinematics in the beginning of the stroke. (*A*) The primary condition: anteromotoric in-phase functional four-wingedness (forewings are leading). (*B*) Advanced anteromotoric in-phase functional four-wingedness. (*C*) Morphologic two-wingedness. (*D*) Functional two-wingedness with coupled fore- and hindwings. (*E*) Posteromotoric in-phase functional four-wingedness (hindwings are leading). The insect is seen from above. Black triangles indicate the upper side of the leading edge of the wings; arrows show the direction of the wings' movement; dotted line indicates rotational flow (vortices).

given rise to the clade of dipterous insects (fig. 3.6C), with the transformation of the hindwings into halteres. This viewpoint is supported by the fact that one of the two patterns of phase relationships between the wings and halteres found in Muscida is typical for the functionally four-winged, in-phase anteromotoric flappers (see sec. 3.9), indicating that the ancestors of dipterous flies did not have linked fore- and hindwings.

The second method to remove the malfunctional interaction between fore and hind starting vortex tubes consists in the coupling of fore- and hindwings into a single flapping surface. Each surface generates one tube (sec. 2.4); hence, the circulation and vortex ring develop without starting deceleration (fig. 3.6D). In this case the magnitude of bound vortex, intensity of vortex ring rotation, and its contribution to the generation of useful forces are evidently increased for the entire stroke period.

The evolution of functional two-wingedness in many taxa is associated with the decrease in relative hindwing size. In some cases the hindwings are completely lost, resulting in the origin of morphologic two-wingedness (fig. 3.6C).

The second strategy results through transfer of the function of the primary vorticity generator from the fore- to the hindwings. It is found in functionally four-winged, in-phase posteromotoric insects. Their stroke cycle begins with opening of the hindwings, leading to earlier formation of the hind starting tubes. Since any vortex is a zone of reduced pressure, the hind starting bubbles induce a flow that prevents the development of vortices behind the forewings (fig. 3.6E). Under a definite intensity of hindwing strokes owing to created rarefaction, formation of vortices after the forewings becomes impossible. Posteromotoric, in-phase four-winged kinematics is more effective than anteromotoric: the weight-specific lift falls 1.5-fold after the phase shift decreases to zero. Locusts use this strategy to regulate flight power output (Wortmann and Zarnack 1993).

The reduced strokes of forewings reduce their thrust and lift relative to the hindwings, whereas drag experienced by the forewings remains roughly the same. Therefore, the aerodynamic effectiveness of flight can be improved through complete transfer of the locomotor function to the hindwings. This path of flight evolution, like the two previously described anteromotoric specializations, becomes complete with the appearance of morphologically two-winged forms (fig. 3.6C).

Thus, the major patterns of the wings' operation and the corresponding patterns of vortex formation, based on the results of comparative flight aerodynamics research, perform adaptations to overcome the negative effect of mutual deceleration of starting vortices in the first moments of stroke cycle. The two strategies to remove interaction between the fore- and hindwings present typical examples of morphofunctional change under long-term stabilizing selection, resulting in decreased energy waste, which is the major consequence of stabilizing selection after Schmalhausen (1949, 1969). It is also quite remarkable that only three (two anteromotoric and a posteromotoric) conceivable opportunities to improve the aerodynamic efficiency of the most primitive anteromotoric, functionally four-winged flying systems exist, and that all three have been realized through the evolution of insect flight.

The suggested explanation does not totally exhaust the problem, because historical transformations of wingbeat kinematics are not confined to the adaptations that result in more mechanically effective aerial propelling agents. The classification of kinematic patterns (fig. 1.8) also includes less efficient varieties, inadaptive from the perspective of flight energy expenditures. These cases, characteristic of some orders, will be considered in the next six sections of this chapter, following Rasnitsyn's phylogeny of insects (Rohdendorf and Rasnitsyn 1980; Rasnitsyn 1997).

3.4.   Grylloid Insects

Most insects included in the infraclass Gryllones are functionally four-winged and posteromotoric (see chap. 1). However, they also include anteromotoric forms, namely stoneflies. Anteromotorism probably characterizes web spinners as well, judging from their larger forewings and mesothorax, as well as from their general similarity to insects in the order Perlida (Martynov 1938), which suggests that the initial wingbeat kinematics of grylloid insects was the anteromotoric functionally four-winged. The possession in stoneflies of features primitive for Gryllones and common to rock crawlers is consistent with the hypothesis that anteromotorism was typical also of this order, normally placed at the root of grylloid insect phylogeny (Rohdendorf and Rasnitsyn 1980). Thus, the entire natural group Perlida + Grylloblattida + Embiida is characterized by the initial kinematics for flying insects.

The origin of Gryllones is connected with the transition to a hidden mode of life (Rohdendorf and Rasnitsyn 1980). This suggests that early grylloids were not exposed to environmental stimuli for flight improvement, and is further supported by the comparatively elusive flight of many contemporary grylloid insects: only some short-horned grasshoppers (Gryllida: Acrididae) can migrate for long distances, and three-fourths of the species in grylloid orders are completely flightless (Dudley 1999).

Despite their posteromotoric kinematics, the flight of orthopteroids can hardly be called advanced. One of the indicators of primitiveness can be a relatively stable kinematics and, hence, small maneuverability. The turn of a flying specimen is accompanied by a roll of the entire body and is carried out within several sequential strokes, whereas the trajectory of the hindwings does not change essentially, but the forewing path acquires asymmetry: the right and left wings work at different angles of attack, their working cycles begin at different moments of time (earlier in the contralateral wing), the wingbeat amplitude is increased at the contralateral and decreased at the ipsilateral side, and the stroke plane of the forewings is turned to the side of the new flight direction (Gettrup and Wilson 1964; Dugard 1967; Baker 1979; Cooter 1979; Kammer 1985; Waldmann and Zarnack 1987, 1988; May et al. 1988). Simultaneously, the insect abdomen curves and the hindleg stretches to the side of the turn (Dugard 1967; Camhi 1970; Baker 1979; Taylor 1981; Arbas 1986; Schmidt and Zarnack 1987; May and Hoy 1990). This turn strategy generates greater lift on the contralateral and greater drag on the ipsilateral side. As will be shown, advanced insect fliers are capable of considerably sharper maneuvers.

The long, independent evolution of grylloid insects resulted in peculiarities of the construction and operation of their wing apparatus. Thus, grylloids possess a specific plaiting of wings that differs greatly from that of other insects (Martynov 1924b, 1938). They are also unique in their use of the lift-control reaction (Gettrup and Wilson 1964). Although this reaction is not always observed in experimental and natural conditions (Zarnack and Wortmann 1989), it has never been found in other insects (e.g., see figs. 2.10 and 2.11); an exception is given only by dipterous flies (sec. 3.9).

The weak flight of grylloid insects may result from the majority of species (primarily the Gryllida, the largest order among Gryllones) feeding on the green parts of plants, food that is low in calories and highly toxic. The digestion of this material requires a bulky gut, which contributes significantly to the total mass of grylloid specimens. Hence, wing loading increases, which requires a higher stroke frequency and enhanced energy consumption, although grylloid insects lack the asynchronous flight musculature necessary for flight at high wingbeat frequencies (Cullen 1974; Dudley 1991b). Relatively heavy digestive systems explain the general rarity of leaf-eating forms among flying animals (Dudley and Vermeij 1992, 1994). This increased mass probably also induced the development of strong jumping hindlimbs in many grylloid insects. A starting jump launches the specimen into flight; these insects are unable to take off without the help of their feet, because takeoff requires kinematics more powerful than in straight flight (sec. 3.1).

3.5. Damselflies and Dragonflies

The Libellulida possess the most diversified kinematics among all the Pterygota. They most typically display antiphase strokes, observed in comparatively small species in all maneuvers and in larger specimens primarily in hovering. This is true for both the zygopteran and anisopteran suborders. Occasionally, they perform flight with only hindwing strokes, while the forewings are immovably stretched out (Rüppell 1989). Sometimes the opposite occurs, and only the forewings operate (Zalessky 1955). Although it is impossible to definitely state the type of kinematics exhibited by ancestors of the Libellulida, anteromotoric functional four-wingedness seems an appropriate candidate and has been found in some species of damselflies (Rüppell 1989).

Characteristically, when dragonflies need to generate enhanced lift (in natural conditions—during takeoff, chasing prey, escaping from danger, accelerating horizontally, transporting prey, stealing insects caught in spider webs, and often in tandem flight in-copula; in experimental conditions— when their wings are artificially loaded or when the specimens are tethered), their wings operate at phase shift inherent for functionally four-winged inphase kinematics (Neville 1960; Alexander 1984; Azuma et al. 1985; Somps and Luttges 1985; Azuma and Watanabe 1988; Rüppell 1989; Wakeling 1993; Wakeling and Ellington 1997b). Hence, posteromotoric kinematics is characterized by a higher aerodynamic power output than antiphase wingbeats.

Counterstrokes are presumably an adaptation to damp vertical oscillations of the body during hunting and to increase flight maneuverability. The predatory mode of life typical of dragonflies requires prey capture in flight, which in turn demands precise location of a rapidly moving target. The periodic oscillations of the body that occur during the stroke cycle in dragonflies and many other insects (see sec. 2.9 and also Ellington [1984], Rüppell [1989], and Dudley [1991a]) interfere with this ability. Apparently antiphase strokes decrease the displacements of the center of mass of hunting dragonflies.

The wings of dragonflies can work independently from one another. A flight mode is even possible in which the wings of one side remain still, whereas those of the opposite side flap. Furthermore, a phenomenon unique among insects was observed in dragonflies, in which the wings of the left and the right sides operate at different frequencies (Rüppell 1989). The increase in the number of degrees of freedom of the wings is directly related to dragonflies possessing a most peculiar morphologic feature: direct flight musculature. This adaptation provides dragonflies an outstanding maneuverability unobtainable by the majority of insects.

For example, in other Scarabaeona, flight direction is changed by turning throughout several consecutive strokes and rolling to the side of the turn (see sec. 3.4). Dragonflies can make turns of this and also another type that is up to three to four times faster (Ryazanova 1966). During a fast turn, a dragonfly changes the inclination of the forewings stroke plane to the side of the new direction, while the stroke plane of the hindwings is inclined to the opposite side. Thus, during a right turn the forewings generate force that turns the head rightward, while the hindwing pair simultaneously turns the abdomen leftward. As a result, within a single stroke or less, the dragonfly rotates around the vertical axis that passes through the border between the meso- and the metathorax (Alexander D. E. 1986). However, note that turns of this type are only possible when the center of mass is located between the hinges of the fore- and hindwings—a feature not typical of other insects.

Having acquired the ability to independently operate each wing, dragonflies began to waste more energy but became much more maneuverable. This kinematics allows them to perform the most diverse maneuvers, to fly sideways and even in reverse direction (Zalessky 1955; Ryazanova 1966; Rüppell 1989). Thus, higher energy waste is compensated by increased hunting effectiveness. Moreover, energy expenditure of the wings as propelling agents is diminished with an increased aspect ratio, because a greater flapping surface area decreases the amount of energy necessary for flapping flight (Weis-Fogh 1964; Heinrich 1974; Pennycuick 1975; Casey et al. 1985).

Thus, dragonflies, having separated early from the root of Scarabaeiformes + Cimiciformes clade, underwent a fundamentally distinct evolution of flight function and acquired two major wing apparatus operating modes: a more powerful in-phase posteromotoric mode with high aerodynamic efficiency, and a less economical but more maneuverable antiphase mode. Selection of this evolutionary route was channeled by a predatory mode of life in open air spaces. Finally, the kinematic lability of dragonflies is described by a set of several highly advanced features and appears as an extreme and narrow specialization on the base of plesiomorphic noncoupled wings. Stabilization of primitive traits is a fairly regular phenomenon in organisms that separated early from the common evolutionary stem of their taxon.

In general, the Libellulida provide a curious case that seems to contradict the concept of symmorphosis, because they possess a flight system apparently constructed in excess of what is normally needed, given their mode of life. This can be seen from the data on stroke amplitude in different flight situations. Increase of amplitude is the most common way to enhance the aerodynamic forces generated by insects (see sec. 2.10). Meanwhile, damselflies and dragonflies fly at low stroke amplitudes (averaging 115° for Zygoptera and 83° for Anisoptera [Rüppell 1989]). The highest value of 150° was recorded in a male *Leucorrhinia rubicunda* (Libellulidae) that was taking off vertically, carrying a female in-copula (Rüppell 1989)! Low stroke amplitude gives an impression that the wing apparatus of dragonflies is capable of much higher force output than normally required. This outstanding ability permits the Libellulida to use their wings in counterstroke kinematics that is evidently ineffective aerodynamically but provides the insects with more efficient hunting behavior. And even during counterstrokes, the amplitude does not exceed 130° (Rüppell 1989).

3.6. Hymenopterans and Rhynchote Insects

The association of these two orders in one section may seem unusual, since rhynchote insects are unrelated to hymenopterans and do not exhibit complete metamorphosis. However, a series of substantial analogies can be traced in the functional and morphologic organization of Vespida and Cimicida wing apparatus.

Both orders are functionally two-winged. Likewise, this state is initial, because forms with uncoupled wings are unknown even in the fossil condition (Rasnitsyn 1980; Rohdendorf and Rasnitsyn 1980). The shape of the flapping surface is also similar: its aspect ratio varies between 1.8 and 4.0 (Danforth 1989) in Hymenoptera and between 1.9 and 3.9 in Heteroptera (Grodnitsky 1996) (fig. 3.5).

The stroke frequency is quite high, considerably exceeding 60 Hz in both orders (fig. 3.3). It is only low in huge belostomatid bugs (total mass 23.4 g, body length 105 mm), at 21–25 Hz (Barber and Pringle 1965), but this is evidently an extraordinary case. In rhynchote insects, the frequency approaches 145 strokes per second (Betts 1986c). Specialized foraging hymenopterans, which frequently transport considerable additional mass during flight, display stroke frequencies of 250–280 Hz (Magnan 1934; Unwin and Corbet 1984). A maximal frequency of 350–370 Hz has been observed in the microscopic wasp *Encarsia* (Weis-Fogh 1973).

The stroke plane is inclined at an angle of 30–60° to the body in all species investigated, except for the jumping plant lice *Amblyrrhina maculata* Low. (Psyllidae). The primary part of the flapping surface is elastic. In addition, the predominant type of wing kinematics in the upper part of the stroke cycle is near clap-and-peel. The wings touch each other by the trailing edges only slightly in the beginning of the stroke. Clap has only been observed in tethered *Allantus* sp. (Tenthredinidae) (Grodnitsky and Morozov 1994). The forewings of the majority of heteropteran and hymenopteran species undergo ventral flexion of their distal parts at the end of the downstroke. The wings of *Urocerus gigas* L. flex slightly but distinctly near the midspan (fig. 3.7A).

Throughout the second half-cycle, the flapping surface is generally concave along several furrows and the line of the wings' coupling (fig. 3.7A). At the same time, in rhynchote insects, the forewing anal field overlaps the costal margin of the hindwing so that the forewing remigio-anal furrow combines with the costal vein of the hindwing. As a result, this furrow and the line of the wings' coupling work as a single deformation line.

In several cases, the loss of independence by the hindwings leads to a decrease in their size and sometimes to their complete reduction, resulting in a morphologically two-winged condition. This state is found in *Mymar* (Mymaridae) and Mymarommatidae from Vespida and Coccoidea among Cimicida. However, note that these are exceptional instances. Absence of the morphologically two-winged condition in flying bugs is presumably caused by the defensive function of their thick cuticle and elytra. Loss of hindwings requires an increase in the stroke frequency of the forewings, which is difficult to achieve with massive elytra. This explains the often considerable dwarfing of the hindwing pair, which nonetheless rarely disappear completely (Popov 1971).

In many higher hymenopterans, hindwing reduction is constrained by their need to take care of progeny, along with the necessity to transport paralyzed prey, nectar, and pollen to provision nests. Hence, the flight of higher hymenopterans evolved in a direction to support load-lifting capacity. These flyers can carry more than twice their own weight (Fabr 1914). Hindwing reduction increases the loading of the forewings, and correspondingly diminishes takeoff ability when carrying additional mass. Note that social

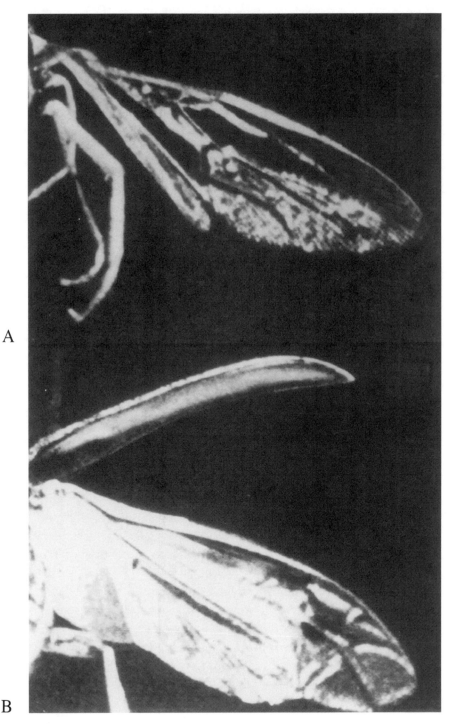

A

B

hymenopterans are characterized by the highest loading found in insects (up to 30–35 N/m^2 and greater), even with four well-developed wings (Magnan 1934; Marden 1987; Dudley and Ellington 1990b). The rarity of morphologically two-winged forms among other higher Hymenoptera (suborder Vespina in general) still remains unexplained.

3.7. Beetles

The majority of beetle species are posteromotoric functionally four-winged insects (fig. 3.8). This condition is associated with the transformation of the forewings into elytra, resulting in transfer of the main propulsive function to the hindwings. Intermediate forms that connect the beetles with more generalized groups of Scarabaeiformes and that indicate the process of formation of coleopteroid appearance are unknown among both recent and fossil forms (Rohdendorf and Rasnitsyn 1980). Therefore, the question of whether posteromotoric flight is primitive for this order remains unanswered and is closely linked to the problem of taxonomy of the ancestors of Scarabaeida.

The stroke frequency of coleopterans is higher than in generalized insects with complete metamorphosis, but noticeably lower than in insects with frequent flaps. In beetles it constitutes from 30 to 40 Hz (fig. 3.3; Magnan 1934; Weis-Fogh 1973) to 100–130 Hz (fig. 3.3; Baker 1965; Weis-Fogh 1973; Grodnitsky and Morozov 1994). The stroke plane of many species is vertical but can be strongly inclined (e.g., in *Cetonia*) (Grodnitsky and Morozov 1994). Wing loading is well above medium insect value (fig. 3.4), owing to the heavy exoskeleton, and stroke amplitude is comparatively high (150–180°), presumably for the same reason.

The elytra of beetles provide a typical example of conflict between two functions, namely flapping flight and protection of the wings and body. In accordance, the kinematics of elytra is strongly modified because they are considerably more massive than regular insect wings.

Schneider (1975) described four flight types in beetles based on differences in the motion of the elytra. In the majority of species, the elytra perform strokes of small (no more than 30–40°) amplitude, and the wings and elytra oscillate almost synchronously (fig. 3.8B–D). In *Cicindela* they are fully

Facing page: Figure 3.7. Deformation of the wings at the beginning of upstroke: (*A*) horntail *U. gigas* L. (Siricidae); (*B*) longhorn beetle *M. sutor* L. (Cerambycidae). Seen from behind.

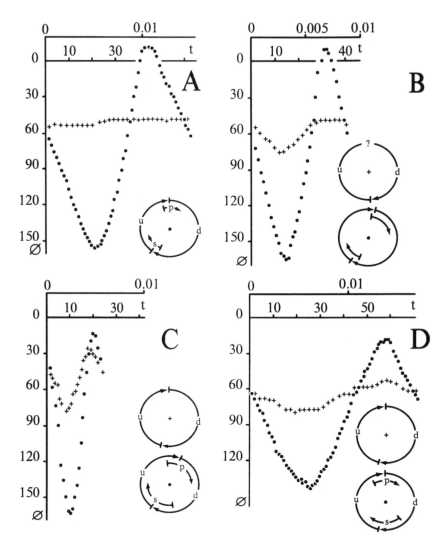

Figure 3.8. Wingbeat kinematics of beetles. (A) tiger beetle *Cicindela restricta* (Carabidae); (B) longhorn beetle *Corymbia rubra* L. (Cerambycidae); (C) tumbling flower beetle *Mordellistena tournieri* Emery. (Mordellidae); (D) longhorn beetle *Monochamus urussovi* Fisch. (Cerambycidae). Designations as in figure 3.2.

stretched and stationary (fig. 3.8A). *Cetonia* only opens the elytra slightly, releasing the wings from under them. *Necrophorus* acts in a similar manner, but was separated by Schneider into a distinct flight type. A flight mode not considered by Schneider also exists: the middle Asian metallic wood-boring beetles *Acmaeodera* and *Acmaeoderella* (Buprestidae) possess fused elytra. Before takeoff they open the elytra like a car hood and in flight hold them raised almost perpendicular to the body (Ivanov V. D., Grodnitsky D. L., unpublished data). These five types of coleopteran flight demonstrate different ways to resolve the contradiction between functions and probably do not exhaust the entire diversity of flight in this very peculiar group.

The major aerodynamic force necessary for flight in beetles is created by the hindwings (Schneider and Hermes 1976). The anal fields of the wings are elastic and in the upper part of their trajectory demonstrate "flexible fling" or peel (see Ellington 1984). Pause in motion (clap) has been observed in several species. In others the wings do not stop between sequential strokes (Grodnitsky and Morozov 1994). In *Cetonia* the left and right wings do not touch each other either over or under the body.

Deformation of the wings in beetles is confined to a smooth convexity during pronation, formation of a gutter in the posteroproximal zone during downstroke, smooth concavity, and (occasionally) bending along the remigio-anal furrow during supination and upstroke. Bending along the concave secondary anal furrow has been observed only in *Cetonia* (Grodnitsky and Morozov 1994). In all cases only the ano-jugal part of the wing is cambered. The remigium remains flat or exhibits a slight gradual profile throughout the entire stroke cycle. The absence of sharp deformations in this zone is quite surprising, since the remigium of beetles contains several secondary longitudinal and transversal furrows that serve for wing plaiting at rest (see sec. 4.5). The presence of these furrows not only fails to generate apomorphic flight deformations of the wings, but instead seems to have an opposite effect, imparting to the remigium a stiffness generally uncharacteristic of this wing region.

The rarity of morphologically two-winged beetles reflects the possession of a heavy external skeleton, and, consequently, the beetles need to generate relatively high lift. Elytra are used for this purpose, but their effectiveness is reduced because of the inability to create thrust. For this reason, most beetles fly slowly in comparison with other insects. The role of the elytra as generators of pure lift is illustrated by the flight characteristics of the flower beetle *Cetonia*. Owing to morphologic two-wingedness, its wings experience extremely high (41.3 N/m^2) loading (Magnan 1934). Hence, this beetle must operate its wings at relatively high (in comparison with other coleopterans) stroke frequency: >100 Hz (Weis-Fogh 1973). The kinematic parameters of beetles therefore provide evidence for enhanced flight power output. Likewise, their wings work in a regime close to the functional limit, since coleopterans have the least additional load-lifting capacity among all the insects (Marden 1987).

3.8. Caddisflies, Moths, and Butterflies

The orders Phryganeida and Papilionida are related and form the monophyletic group Amphiesmenoptera, of a rank intermediate between order and superorder. Caddisflies and moths possess many similar features of wing morphology and function. At the same time, the functional and morphologic diversity of modifications of the wing apparatus is noticeably higher in the Amphiesmenoptera than within any large Scarabaeona order. The most primitive recent Amphiesmenoptera are anteromotoric functionally four-winged forms. Their flight kinematics does not differ substantially from that of Myrmeleontida and Panorpida, which include caddisflies Rhyacophilidae (Ivanov 1985a), primary moths Micropterigidae and Eriocraniidae, and ghost moths Hepialidae (Grodnitsky and Kozlov 1991). They differ from neuropterans and scorpion flies primarily because at the end of downstroke, the wings of many species undergo ventral flexion of the distal part. As a rule the distal third region of the wing that bends down demonstrates peculiar trapezia-like deformation (fig. 3.9).

All Amphiesmenoptera, except the clearwing moth *Aegeria,* have a vertical stroke plane. The fore- and hindwings of caddisflies and moths operate in flight as an entire lifting surface, even when special coupling mechanisms are absent, as for example in butterflies. The trajectories of both wings are therefore almost identical, although the amplitude of the hind pair oscillations is slightly increased. The construction of coupling mechanisms displays considerable morphologic diversity in different groups (Ivanov 1989; Borror et al. 1992), indicating the independence of their origins and the opportunity of an organism to obtain an identical functional result through different morphologic adaptations.

The stroke frequency of caddisflies is quite uniform and constitutes from 16 to 50 Hz (fig. 3.3). In lepidopterans the wingbeat frequency is considerably more diverse (fig. 3.3), owing to the presence in the Papilionida of two alternative adaptive strategies related to either ecto- or endothermic regu-

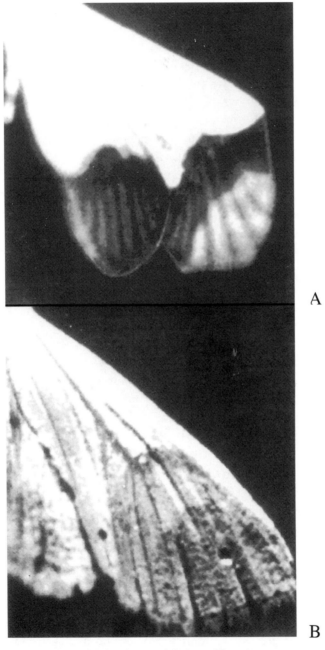

A

B

Figure 3.9. Abrupt distal ventral flexion of the wings
during their deceleration in the bottom of trajectory:
(*A*) caddisfly *Apatania* sp. (Limnephilidae); (*B*) cotton moth
Heliothis armigera Hbn. (Noctuidae). Seen from behind.

lation of body temperature (see, e.g., Hochachka and Somero [1973], Casey [1981], and Heinrich [1981, 1993]). Ectothermic animals gain most of their heat from the surrounding environment by basking in the sunlight. Endothermic organisms heat themselves through intensive muscle contraction. In addition, endothermic lepidopterans are, as a rule, nocturnal (Heath and Adams 1965; Heinrich 1971, 1993), whereas ectothermic species are diurnal forms (Vielmetter 1958; Clench 1966; Watt 1968).

In large lepidopterans that fly well, the specific thermoregulatory strategy defined two major evolutionary ways of wing transformation: broadening of the wings followed by reduction of their flight stroke frequency, and narrowing of the flapping surface under simultaneous increase of frequency (fig. 3.10).

The endothermic temperature regulation observed in representatives of the superfamilies Sphingoidea, Noctuoidea, and Notodontoidea requires high stroke frequency. Thus, the progressive evolution of flight function indicated by an increase in flight range, rapidity, and maneuverability in endothermic moths was achieved in parallel with narrowing of the wing surface at a relatively high (60–85 Hz) stroke frequency.

Improvement of flight characteristics in diurnal ectothermic butterflies (Hesperioidea, Papilionoidea) was accompanied by widening of both wing pairs, because they play the role of heat accumulators (Wasserthal 1975). Broadening of the wing surface accomplishes more effective ectothermic heating but, at the same time, limits stroke frequency, which in large Pieridae and Papilionidae decreases to 4–8 Hz.

The use of solar radiation as a source of heat allowed butterflies to reduce the energy consumption of the wing apparatus as a locomotor system. On the other hand, the adaptive value of endothermic temperature regulation is also evident: the development of a powerful thoracic musculature capable of prolonged work at high contractile frequency enhanced the load-lifting capacity of the wing apparatus of moths, permitting increased total body mass, enhanced fecundity, and storage of fat and other reserve compounds.

Broad-winged forms may also be found among crepuscular and nocturnal lepidopterans; these include giant silkworm moths, many inchworm moths, and related families. Giant silkworm moths Saturniidae are endothermic lepidopterans that engage in preflight warm-up (Hanegan and Heath 1970a, 1970b) with temperature indices extremely similar to analogous characteristics of hawkmoths (Bartholomew et al. 1981). The saturniid adults do not eat and have reduced mouthparts, so the moths have limited reserves of ener-

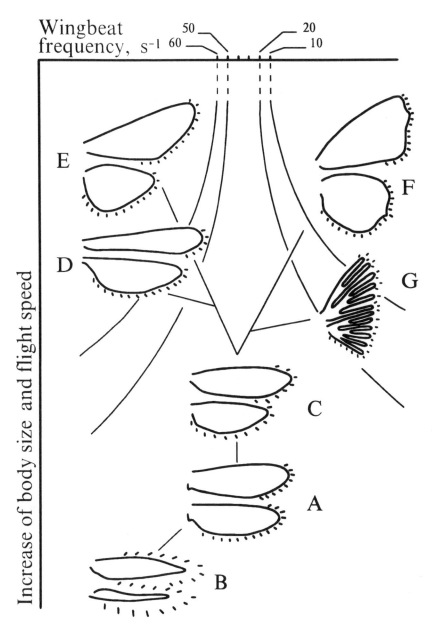

Figure 3.10. Morphologic and functional evolution of the wings in moths and butterflies: (A) starting condition; (B) feather-like wings; (C) the majority of microlepidopterans; (D) narrow wings of clearwing moths; (E) narrow wings of hawkmoths; (F) broad wings; (G) split wings of plume moths.

getic substrate in their bodies. This probably contributed to the secondarily increased wing surface observed in saturniids, since broadened wings decrease wing loading and consequently reduce energy expenditures during flight (Dorsett 1962; Weis-Fogh 1964). The heat produced during the preflight warm-up is retained through the insulation provided by a dense scale covering. In addition, some species (e.g., *Aglia tau* L.) have become secondarily diurnal.

Inchworm moths Geometridae are apparently neither endo- nor ectothermic insects, because the temperature of the thoracic muscles of *Operophtera bruceata* Hulst. and *Alsophila pometaria* Harris, flying in an air temperature of −3°C, equals 0° (Heinrich and Mommsen 1985). The biochemical and tissue adaptations that permit "cold" muscles to work are comprehensively discussed by Heinrich (1993).

Some amphiesmenopteran groups are characterized by feather-like wings with a long marginal fringe of scales (fig. 3.10*B*), and by wings split into two or three feather-like lobes (fig. 3.10*G*) with a hairy fringe overlapping the gaps between the lobes. Contrary to the early assumption made by Rohdendorf (1949), this construction is not associated with flight peculiarities: wingbeat kinematics in these groups is not significantly different from that of caddisflies and lepidopterans with whole wings (Norberg 1972c; Ellington 1984; Kozlov et al. 1986). In fact, a thick boundary layer does not allow air to pass through the fringe, so feather-winged and split-winged moths fly in the same manner as other Papilionida.

The formation of feather-like wings is a reaction to the decrease in the absolute size of specimens. In tiny species the fringe is relatively stronger than in large insects. In addition, it is lighter and thus of greater adaptive value than a solid wing. A split flapping plane may be associated with wing plaiting (Wasserthal 1974), because splits along furrows prevent the appearance of objectional deformations of the wing membrane when the wings are being folded at rest. Indeed, more or less deep cuts often mark the place of a furrow influx into the wing edge. However, many-plumed moths Alucitidae do not fold up their wings at plaiting, so this is an equivocal explanation.

Development of thermoregulatory ability in lepidopterans is closely related to the most distinct morphologic feature of the order, namely the presence of a covering of the flattened hairs called *scales*. The covering is a good heat insulator (Clench 1966; Heinrich 1970; Heinrich and Bartholomew 1971), and its structure reflects the peculiarities of moth thermoregulatory physiology (see sec. 4.8.3). Furthermore, the covering contributes considerable mass to the wings. For example, in Heliconiidae butterflies scales constitute up to one-third of the total wing mass (Dudley 1999), whereas in noctuid, lasiocampid, and other large moths the covering can be much heavier. This additional mass constrains the increase in stroke frequency in Papilionida. The potential danger of overheating from intense muscle contractions provides another source of constraint on wingbeat frequency. Consequently, no lepidopterans have a wingbeat frequency above 70–90 Hz (fig. 3.3; Magnan 1934; Weis-Fogh 1973; Bartholomew and Casey 1978), and flappers with the highest wingbeat frequencies (e.g., clearwing moths) have lost scales on their wings. The prohibition of high stroke frequency agrees well with the absence of asynchronous thoracic musculature in lepidopterans (Dudley 1991b) and the exceptional rarity of morphologically two-winged species among moths. The specific character of the flight performance of Papilionida is defined by the numerous forms that have extremely low frequency, aspect ratio, and wing loading (figs. 3.3–3.5).

An important peculiarity of lepidopteran kinematics is that the wings constantly and smoothly change their geometric angle of attack throughout the stroke cycle. The stroke cycle does not contain periods when this angle is constant (Grodnitsky and Morozov 1994). Kinematics and deformation of the wings in the upper part of their trajectory can be different. Amphiesmenopterans principally exhibit the flexible modification of fling termed *peel* (Ellington 1984). The wings open gradually during peel, like two sheets of thin paper. In addition, many of the studied specimens had no period in their wingbeat cycle, when the left and right wings completely adjoined each other over the body. Thus, the wings often fail to pause in the upper part of the trajectory, and their motion is not interrupted between the two consecutive strokes. This motion was named *near clap-and-peel* (Ellington 1984).

Wing deformation at the end of the downstroke and during supination is even more diverse (Grodnitsky and Morozov 1994). In many species typical tip ventral flexion is observed accompanied by a trapezia-like deformation (fig. 3.9), persisting within the main part of upstroke. In addition, smooth wing flexion can take place, often accompanied by downward bending of the entire costal margin. Characteristically, sharp flexion of the distal third is primarily seen in generalized caddisflies and butterflies with stroke frequency medium for the group (20–40 Hz). This type of deformation was not observed in specialized lepidopterans (both narrow- and broad-winged); their wings bend smoothly instead. Sharp bending along furrows is also less pronounced in these insects (Grodnitsky and Morozov 1994).

Many broad-winged species that fold up their wings along the abdomen at rest acquired additional furrows, so that caddisflies usually have seven furrows on the hindwing, whereas butterflies generally have five (sec. 4.2.4; [Grodnitsky and Kozlov 1990b]). Deformation of an amphiesmenopteran wing along the secondary concave furrows that serve for plaiting of the wings in the rest position has never been observed. Surprisingly, the presence of completely developed remigial and remigio-anal furrows in many specimens does not produce additional flexion of the wings throughout the entire second half-cycle (Grodnitsky and Morozov 1994).

3.9. Dipterous Flies

The most primitive wing apparatus among the Muscida is found in crane flies of the suborder Tipulomorpha. Their venation is believed to be close to ancestral (Hennig 1981; Wootton and Ennos 1989; Shcherbakov et al. 1995), their stroke frequency varies from 40 to 60 Hz (Magnan 1934; Zalessky 1955; Weis-Fogh 1973; Ellington 1984; Brodsky 1985; Ennos 1989a), and their wing loading is low (4 to 5 N/m^2) (Ennos 1989a). Additionally, crane flies possess an obviously secondary wing planform, the longest within the order (aspect ratio = 4.0–6.0) (Ellington 1984; Ennos 1989a).

The functional and morphologic features of the flying system in dipterous flies evolved from the initial condition toward increased stroke frequency, with an upper limit not yet determined. Although I have never observed a frequency over 150 Hz, published data indicate that stroke frequency can reach 340 Hz and possibly higher in flower flies (Weis-Fogh 1973). The stroke plane in all dipterous flies is inclined to the longitudinal axis of the body at an angle of 30–60°.

The hindwings of dipterous flies are transformed into halteres and have changed flight function for aeronavigational purposes (Pringle 1963; Nalbach and Hengstenberg 1986). Synchronization of the strokes of wings and halteres in the crane fly *T. oleracea* is the same as in functionally four-winged insects: the halteres move slightly behind but perform strokes of noticeably greater amplitude (fig. 3.11*A*). In robber flies *Leptogaster*, halteres also have a larger stroke angle, but move in antiphase to the wings (fig. 3.11*B*). In all likelihood, because halteres play the role of organs of perception that work like gyroscopes, they can operate in different modes. The functional role of such kinematic modifications is still unclear.

The development of the ability to flap the wings at high frequency was

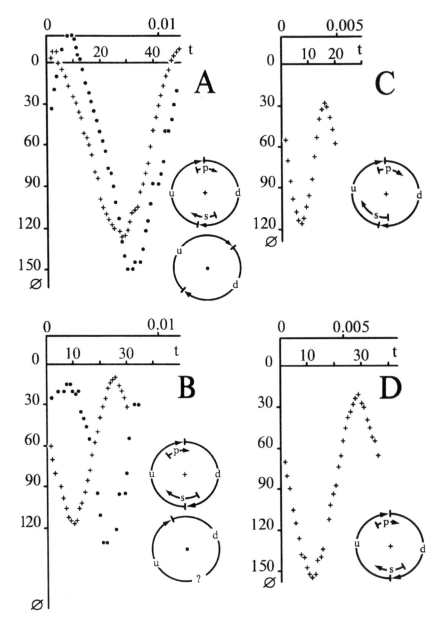

Figure 3.11. Wingbeat kinematics of dipterous flies: (*A*) crane fly *Tipula oleracea* L. (Tipulidae); (*B*) grass fly *Leptogaster cylindrica* DeGeer (Asilidae); (*C*) flower fly *Sphaerophoria taeniata* Mg. (Syrphidae); (*D*) dipterous fly *Tephrochlaena* sp. (Helomyzidae). Designations as in figure 3.2.

accompanied by an increase in wing mobility relative to its longitudinal axis, favored by a narrow wing hinge—the most typical morphologic feature of Muscida that enables more effective use of the aerodynamic effect of fast rotation of the wing around its hinge-tip axis (Dickinson et al. 1993). The formation of narrow wing hinges may also have resulted because propeller thrust loses effectiveness if it is dissipated against the body (Vogel 1965). Thus, stalked wings generate thrust farther from the body than wings in other insects. This is especially important for insects that create a considerable part of their aerodynamic force with the help of rotational movements of the wings.

The wing's angle of attack is under at least partly automatic control (Ennos 1987, 1989b; Wootton and Ennos 1989), in which the left and right wings can work in different regimes, or gears, according to the terminology of Pfau (1973, 1977, 1985, 1987; Nachtigall 1985; Nalbach 1988, 1989; Wisser 1988). Also, in dipterous flies one of the wings can perform complete strokes, while the other remains still (Nachtigall 1985).

The wings of dipterous flies never clap together in the upper part of their trajectory. The wingbeat path is reduced in both the upper and lower parts (fig. 3.11). Finally, as a rule, near clap-and-fling is observed. Often the wings do not adjoin each other over the body at all. Wing deformation is always gradual, without sharp bending, although some species (mainly generalized) have well-developed furrows (fig. 4.3). Sharp tip ventral flexion can be observed only in crane flies, not in more advanced representatives of the order (Grodnitsky and Morozov 1994). The wing tips of higher flies (suborder Brachycera) never bend ventrally; instead, their wings can bend gradually along the entire costal edge or closer to the hinge (Ennos 1989c).

The basal part of the anal wing region in many Muscida Brachycera is deeply cut out, separated from the rest of the wing by a broad concave furrow, and is known as the alula (fig. 4.3C–H). During the first half-stroke, the alula remains within the wing plane. During supination, it is bent dorsally and kept approximately perpendicular to the wing. The alula remains in this position throughout the greater part of upstroke and returns to its initial position only with the beginning of pronation (Grodnitsky and Morozov 1994). Thus, the alula can be considered as an adaptation to regulate the aerodynamically effective size of the wings, enlarging it in the first and decreasing it in the second half-stroke. Nevertheless, owing to its small size, the total impact from the alula cannot be great.

The functional and morphologic peculiarities of the wing apparatus (navigatory role of halteres, narrow wing hinge, relative independence of left and right wings, automatic control of the angle of attack, ability of "gear change" and of the aerodynamically active wing area regulation) at high stroke frequency enabled swift maneuverable flight typical of the majority of dipterous flies, which are able to fly sideways and backward, perform the most diverse maneuvers with unchanged body orientation or, by contrast, turn within several milliseconds (Collett and Land 1975; Wagner 1982; Kammer 1985; Nachtigall 1985). Another exceptional feature of dipteran flight is the capacity to regulate thrust, lift, and torque independently (Vogel 1966; Götz 1968; Spüler and Heide 1978; Blondeau 1981)—a quality absent in other insects, in which vertical and horizontal components of the resultant aerodynamic force are generally negatively correlated, depending on the angle between the stroke plane and the horizon (figs. 2.10 and 2.11).

The appearance of morphologically two-winged flight, providing such pronounced benefits, is prohibited in other orders by specific functional and morphologic features (see sec. 3.4 and 3.6–3.8) that are not present in the Muscida. At the same time, it remains obscure why an intermediate state of wing apparatus (fig. 3.6B) is found only in the fossil scorpion flies Liassophilidae and Permotanyderidae, neither of which can be considered ancestral to the Muscida (Novokshonov 1993; Shcherbakov et al. 1995), and why it does not occur among recent forms, not considering Nemopteridae (Myrmeleontida) and some Papilionidae, flying only with the help of forewings (chap. 1). This query was posed by V. V. Zherikhin (private communication), but the available information does not provide an answer.

3.10. Functional Assessment of Some Elements of Wing Kinematics and Deformation

Having observed the diversity of wingbeat kinematics found in different lines of insect phylogeny, it is interesting to assess peculiarities of wing kinematics and deformation from the viewpoint of the aerodynamic phenomena that accompany a generalized wingbeat cycle, which were summarized in sections 2.7 and 2.8.

Fling undoubtedly generates useful force. Its effectiveness is a subtle function of the absolute size of an insect, although it can slightly increase with diminution of the flier (Lighthill 1978). However, its contribution to the total balance of forces in medium-sized and large insects is probably not great. The impact of fling could be quite high if the airflow separated from the wing leading edge and formed a corresponding vortex structure. But visualization experiments on flow around living flying insects indicate that separation and vortex formation near the leading edge of the flapping surface take place only at rather high Reynolds numbers and are absent in the range of speeds at which the majority of insects fly in nature (sec. 2.7).

Likewise, fling has only a minor effect in the flight of most insects. In tethered flight, it is usually incomplete and is performed as either "near fling" or "near peel," or it can be absolutely absent. In free flight the wings also do not necessarily touch each other over the body (Wootton 1992). This is primarily characteristic of medium-sized and large excellent fliers: dipterous flies, dragonflies, flower beetles. Therefore, the rarity of full fling cannot be ascribed to inadequate experimental conditions, because in this case experimental data coincide well with observations.

Thus many insects do not use fling while flying. This is probably because the implementation of fling is always associated with a pause in the wing motion, inevitably resulting in a transient decrease in generated aerodynamic force. Therefore, a good alternative would be kinematics without pause, wherein the starting bubbles merge somewhat later, but time is saved by means of stroke cycle continuity. Both patterns can be observed in a single insect species. For example, butterflies demonstrate typical fling during the first stroke at takeoff but often avoid fling in subsequent flight, not stopping their wings in the upper point of their trajectory (Brackenbury 1992). Also, neither full fling nor peel occur in insects with high wingbeat frequency— rhynchote insects, hymenopterans, and dipterous flies (sec. 3.6 and 3.9).

The contribution from fling to the generation of useful forces can be higher in tiny insects, because at low Reynolds numbers the role of viscous effects increases, leading to diminished effectiveness of the mechanisms related to vortex formation. Hence, the total part of force, originating from fling, can increase (Zanker 1990a). Fling is followed by the swinging-edge mechanism, which consists in a quick change of the angle of attack, occurring immediately after the trailing edges depart. According to Nachtigall's (1980) assumption, this mechanism reinforces vortex formation over the wing. Thus, the "*Kantenschwung*" is likewise not a separate aerodynamic mechanism. Indeed, at this moment formation of starting vortices over the wing dorsal surface is observed, and thus the Kantenschwung must intensify the vortices, increasing circulation as well as the aerodynamic effect associated with the influence of the starting vortex. It is difficult to assess any positive effect gained from this mechanism, because any works on its numerical evaluation are absent.

Having completed the Kantenschwung, the wings move downward at a high angle of attack. Generation of useful forces in this period results from circulation flow around the flapping surface and presumably from the presence over the body of a large U-shaped vortex closed to the forewing tips. The ratio of contribution of these two factors to total impulse is presently unknown; the vortex influence is probably stronger in insects with low advance ratio. The formation of a convex profile in insects with relatively broad wings can be adaptively significant, because in this case the wing surface comes closer to the zone of intensive vortex motion and therefore experiences more significant aerodynamic loading (Belotserkovsky and Nisht 1978).

Wings create negative lift during their upward motion, owing to the appearance of negative circulation and the formation of stopping vortices below them. The negative effect can be diminished by means of comparatively low angles of attack typical of the upstroke. To achieve this, the wing chord maximally approaches the stroke plane. In species with a narrow hinge (primarily Muscida), this is achieved by a strong twisting of the wing hinges. Insects with coupled or relatively broad wings, for which hinge twisting is prohibited, demonstrate bending along furrows and the line of the wings' coupling. This explains the absence of well-developed furrows on wings with a narrow hinge and their presence in forms with a comparatively broad flapping surface.

During supination, in many species, ventral flexion of distal parts of the wings takes place. This deformation pattern is thought to serve as an adaptation for damping out wing inertia at the point of reversal movement (Wootton 1981, 1990, 1996; Brodsky and Ivanov 1983b; Ennos 1989c). Contrary to Brodsky's (1985) assumptions, tip flexion does not induce new vortices. Since the flexion persists during a significant part of the following half-cycle, it is possible that, accompanied by trapezia-like deformation (whereby it is expressed), it also serves to decrease the aerodynamically effective size of the flapping surface, thereby diminishing the effect of nonuseful force generated during the second half-cycle. Note that change of effective wing size is widely used by birds and partly by insects (Wootton 1990).

Ventral flexion occurs to a varying extent in different groups and can be associated with the wings' inertia dependent on their mass and motion speed, that is, stroke frequency and amplitude. In slowly flapping insects with relatively light wings (*Chrysopa, Panorpa*), flexion is absent or weak. Increase in wing mass related, for example, to the development of a thick scale or hair covering (Lepidoptera, Trichoptera) augments flexion. Flexion is gradual in species with low frequency (*Leptidea*), and abrupt in those that have rather frequent flaps (*Apatania, Heliothis*).

In insects with high wingbeat frequency, the zone of flexion (nodal line) is shifted basally: to the midspan (*Urocerus*) or closer to the hinge, as in many dipterous flies Brachycera. In high-frequency insects, the flexion is less sharp than in moths and caddisflies; this is supposedly related to strong accelerations experienced by frequently flapping wings. In such a situation, sharp deformations could be dangerous and destroy the wings. Ventral flexion is completely reduced in many species with high stroke frequency. In these groups inertia is damped out at stroke reversal by means of fast pronation and supination (Ennos 1988). A sclerotized, broad flapping surface operating at high stroke frequency is found only in hemipterans. In all likelihood, this particular combination of functional and morphologic circumstances conditioned the peculiar construction of their wings and formation of the membrane—a part of the forewing, having reduced mass and undergoing ventral flexion (Betts 1986c). Notably a discrepancy exists between the membrane morphology and deformation: the line of flexion does not necessarily coincide with the border of the corium and membrane and often passes distad (Wootton 1996). Aside from the flight function, wing flexion is used by flies while cleaning their wings (Zanker 1990a).

3.11. Conclusions

The origin of insect flight constitutes a characteristic case of generation of a new organismal construction, a new *Bauplan*, an evolutionary innovation caused by the collaborative action of morphogenetic and selective forces. Dogel' (1954, 1981) suggested the most general path for origination of novel constructions in his concept of polymerization and oligomerization in evolution: Primordial elements, from which the novelty is then assembled, initially appear as numerous serial homologues, that is, as typical manifestations of ontogenetic regularities. Next, natural selection enhances some of them and suppresses others. Kokshaysky (1974, 1980) called this process *concentration of a function*. Thus, a new organ appears in the part of the organism's body where it is needed. In other words, morphogenesis offers numerous similar minor traits that are then differentiated by natural selection and transformed into general adaptations.

Initially, the terrestrial hexapod ancestors of winged insects had serial, lateral cuticular thickenings on each segment nota. Intensive locomotion is always adaptive in the general sense, enabling prey or mate capture and escape from predators. Hence, the early ability to accumulate solar radiation to intensify muscle contraction is predictable. To accumulate heat that comes from above, any widening of the nota is advantageous, but especially of those located close to the locomotor muscles. Notal cuticular thickenings, therefore, were enlarged and became what we call paranota, most pronounced on the thoracic segments. At this stage paranota of smaller size were situated on the abdominal segments. They were potentially useful in increasing the rate of digestion and maturation of reproductive products through heat accumulation. The thoracic paranota, primarily used for regulation of body temperature, next accepted the ability for gliding, then for flight, and finally became wings. In parallel, insects developed bulky thoracic muscles capable of intensive contractions and inevitable heat generation. The internal heat was further used to warm the rest of the body; thus, abdominal paranota became unnecessary and in most species were strongly reduced.

The first winged insects had four wings that flapped synchronously with a small phase shift, with the forewings leading the hindwings. This kinematic condition was characterized by starting vortex generation independently by each wing in the beginning of each stroke cycle. The vortices interfered with one another, preventing rapid growth of aerodynamic force. The primary functional result insects achieved during early flight evolution was removal of this interference either by reducing one of the two wing pairs or by uniting the fore- and hindwings into two couples, so that only one starting vortex was generated per stroke. The principal types of strokes originated as a result.

An interesting and important correlation can be traced between the constructions of insect flying and feeding systems. Insects with the most primitive anteromotoric functionally four-winged flight (scorpion flies, lacewings, primitive moths, stoneflies, snakeflies and alderflies) have mouthparts of the primitive mandibulate type. All the insect groups specialized on posteromotoric flight (dragonflies, grasshoppers, crickets, mantids, cockroaches, earwigs, and beetles) retain the mandibulate mouthparts. At the same time, all advanced anteromotoric flyers (dipterous flies, moths, butterflies, wasps, bees, and bugs) possess mouthparts of the haustellate type. The only exception are mayflies, which are anteromotoric insects without a haustellum. However, mayflies also do not have mandibulate mouthparts, because their imago do not feed.

The described correlation probably has an explanation in functional and ecologic terms. The most usual reason for the origin of posteromotorism is the participation of the forewings in the defensive function: as the forewings become thicker, stiffer, and therefore heavier, the flapping flight function becomes a responsibility of the hindwings. The defensive properties of the forewings are thought to correlate with the hidden life mode of imago dwelling in the litter of prehistoric forests. Such an ecology is assumed for primitive Scarabaeida and Gryllones, which supposedly lived inside substrate (Rohdendorf and Rasnitsyn 1980). In all likelihood, they were carnivorous and detritophagous insects; this mode of life agrees well with the mandibulate mouthparts. By contrast, the ancestors of anteromotoric insects probably inhabited tall vegetation. The open environment stimulated further evolution of active flight; thus functional and morphologic two-wingedness appeared. Moreover, the origination of different kinds of haustellum was stimulated by the availability of unoccupied vast food sources such as the liquid contents of plant stems and generative organs and, later, animal blood. The presented explanation is, of course, very preliminary.

The further evolutionary improvement of flight function included an increase in efficiency, power output or energy effectiveness, maneuverability, and control mechanisms of flight. These parameters could change quite independently of one another. Comparison of the data on wing motion and vortex dynamics allows an explanation of the main diversity of kinematic phenomena in terms of adaptation. Adaptive significance can be ascribed to aspects of stroke asymmetry, type of phase relationships between fore- and hindwings, fling, and swinging-edge mechanisms. Simultaneously, different flight functional peculiarities obtain explanation for separate taxa, for example, typical frequency and amplitude of strokes, flight speed, and capability (or incapability) to perform specific maneuvers and to take off from the ground. Also, some morphologic features become understandable, namely the total number of wings, relative size of fore- and hindwing pairs, width of the flapping surface (in Papilionida) and wing hinges, and such apomorphic ordinal traits as coupling adaptations and the alula in the Muscida.

On the other hand, few constructional features of the wings can be interpreted in this manner. The problem of evolutionary correlation between structure and function is much more complex. It is considered in detail, using examples of endopterygote insect wing morphology, in chapter 4.

4 | Problems of Endopterygote Insect Wing Functional Morphology

The title of this chapter reflects the complexity inherent in any explanation of the morphology of organisms. There are few cases in which the cause of the evolutionary formation of specific morphologic features of insect wings is clear. Several such examples were given in chapter 3, specifically the features that describe the common principles of wing apparatus construction: the total number and relative size of the wings. Meanwhile, published literature contains various opinions on the functions of peculiar structures, expressed in the form of hypotheses. Direct verification of these assumptions requires special equipment and is often, in principle, unrealizable. Nevertheless, verification is possible to some extent through the comparative analysis of the wing morphology of different groups of insects demonstrating homoplasies, that is, features of secondary similarity that have independently originated on different evolutionary pathways.

Classification of secondary similarities presented in the appendix under "Basic Definitions" (see tab. A1) is based on the reasons causing the origination of homoplasies. However, concrete evolutionary basis for the origin of secondary similarity in different organisms is often obscure. These characters are therefore usually classified not by the reason but rather by the degree of relatedness of the organisms. If the compared species are nonrelated (e.g., belong to different orders) the similarity is ascribed to natural selection, which causes evolution of similar functions, resulting in the origin of convergent (analogous) structures (Darwin 1872; Plate 1928; Schmalhausen 1969; Bliakher 1976; Mednikov 1980; Simpson 1980; Vorobjova 1980; Menner and Makridin 1988). Functional identity is traditionally considered an adequate explanation for identity of form. Hence, we can confirm or reject a functional interpretation of a given structure by assessing the reliability of a similar functional interpretation of the same structure in an unrelated taxon. This method of indirect assay is used in the discussion of the material presented in this chapter.

The terminology for veins and furrows will initially be introduced, then the particular wing morphologies of all endopterygote insect main orders will be discussed and assessed from the perspective of insect wing comparative functional morphology, after considering all currently available functional interpretations of given structural features. The appraisal will not be restricted to any particular group of insects; a broader approach will prevent the confusing speculations often made when the morphology of only a single taxon is considered.

Insects with complete metamorphosis account for more than 655,000 known species (Borror et al. 1992). A complete overview of wing construction of this entire group is consequently impossible, and the focus of research must be limited. In this respect, venation was studied in greatest detail in caddisflies, moths, and butterflies, since these insects give long series of forms intermediate between primitive and advanced, unlike other large orders of Scarabaeiformes that first appeared in fossil history in their typical form.

Within the discussion presented subsequently, the principal attention will be paid to furrows rather than veins. Furrow topography was studied more thoroughly in the Muscida, Vespida, Phryganeida, and Papilionida (Grodnitsky 1991a, 1991b; Grodnitsky and Morozov 1994). The remaining large holometabolan order, the beetles, will be represented by only a dozen species, owing to the current situation regarding beetle wing comparative morphology. Two large articles have recently been published (Kukalova-Peck and Lawrence 1993; Zherikhin and Gratshev 1995). Both articles are extremely rich in original material. However, they also contain a vast number of mutual contradictions, so that a special examination seems to be necessary for appropriate comprehension, the more so because the difference in opinions expressed on beetle wing morphology is not closely related to the context of this book. I used data on comparative morphology as no more than a tool to assess functional hypotheses, and an intensive analysis of structural diversity in all the subtaxa of Scarabaeiformes was not initially planned. The wing covering of lepidopterans is considered an interesting and poorly studied example of a peculiar wing part.

4.1. Topology of Veins and Furrows

Each vein, as a rule, has a characteristic location relative to the wing membrane in the basal part of the wing. Thus, all the veins are divided into upper (radius, generally designated as R; radial sector, RS), lower (subcosta, Sc; media, M; cubitus posterior, CuP), and neutral (the remainder) (Rohdendorf 1949; Emelianov 1977). An exception is presented by the anterior branch of the cubitus (CuA) of caddisflies, which occupies an upper location on the forewings and a lower on the hindwings. This state of the CuA is observed in many taxonomically nonrelated insects, as well as in representatives of the order Paoliida, thought to be the most probable ancestor of all Scarabaeiformes (Rasnitsyn 1980). In all likelihood, the upper location of the anterocubital vein on the forewing and lower on the hindwing is primitive for winged insects. At a distance from the axillary sclerites, the type of vein convexity changes differently in different species; hence, it is impossible to trace a regularity in their arrangement, as was observed by Zalessky (1943). For example, in *Chaetopteryx villosa* F. (Phryganeida: Limnephilidae), all the veins of the distal wing part are raised above the membrane, whereas in *Narosoideus flavidorsalis* Stgr. (Papilionida: Limacodidae), they are depressed instead (Grodnitsky 1991a).

Individual veins can be strengthened in various ways. Wing strength is primarily increased through the formation of a secondary relief of the wing membrane. In dipterous flies, beetles, and hymenopterans, this can be achieved by microscopic corrugation along the wing margin (Bocharova-Messner 1979a: figs. 172, 173, 180).

Secondary gutters can be large and are usually located along the main veins. In some cases they become veinlike because of secondary sclerotization and thickening of the membrane. The best example is given by the false vein "*vena spuria*" of flower flies (Syrphidae). A similar formation is found in caddisflies—a characteristic cuticular thickening that goes along the vein for its entire length. This structure is located anterior to the CuA on the hindwing of many caddisflies (Hydropsychidae, Limnephilidae, Molannidae, Leptoceridae) (fig. 4.1*A*, *C*) and some moths (e.g., *Illiberis sinensis* Walk., Zygaenidae). Sukacheva (1976) and Rasnitsyn (1980) designate the precubital thickening as a sclerotized furrow. It is difficult to agree with this viewpoint because no furrow occurs in this region in either primitive or advanced wings. Novokshonov (1992, 1994) described the vein M5, also located in front

Figure 4.1. Hindwings of caddisflies: (*A*) *Hydropsyche nevae* Kol. (Hydropsychidae); (*B*) *A. wallengreni* McL. (Limnephilidae); (*C*) *Anabolia laevis* Zett. (Limnephilidae). Dotted line, furrows; dashed line, weakened veins; crosses, precubital thickening. For designations of veins see text. aju(+), ano-jugal furrow. Plus after the furrow name indicates the furrow is convex; parentheses indicate the furrow participates in wing folding. Scale 2 mm.

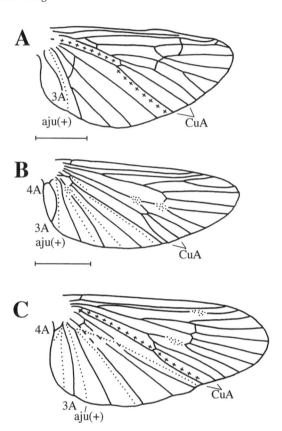

of the CuA parallel to it, on the wings of fossil caddisflies and scorpion flies. This vein is also present in other fossil insects (Rasnitsyn 1980). However, M5 and the precubital thickening are quite distinct, because in the Permian neuropteran *Permopsychops saurensis* Novok. et Viles., they have different basal parts: M5 comes from the medial vein, and the thickening, as in recent caddisflies, begins in the membrane (V. G. Novokshonov, personal communication).

Veins speckled with numerous transverse incisions are present on the wings of some insects. Bocharova-Messner (1979a) suggested that the incised veins function like breather tubes: when the wing is flexed, the transverse corrugation prevents constriction of blood vessels, trachea, and nerves within the vein. Such a vein structure is extremely valuable for insects that fold up their wings at rest; it has an evident adaptive significance in beetles (*Cicindela, Trichius, Oxyporus,* and many others (fig. 4.2*A, B, D*). However, the corrugation of longitudinal and cross veins also occurs in species that never fold up

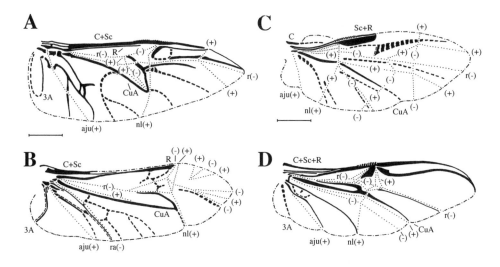

Figure 4.2. Wing morphology of beetles: (*A*) *C. restricta* (Carabidae); (*B*) *Corymbia rubra* L. (Cerambycidae); (*C*) *Oxyporus* sp. (Staphylinidae); (*D*) *T. fasciatus* L. (Scarabaeidae). Dot-and-dash, marginal zone where C is lost. For designations of veins see text. Primary furrows: r, remigial; ra, remigio-anal; aju, ano-jugal. A plus or minus after a furrow name indicates the furrow is convex or concave. A plus or minus in parentheses indicates the furrow participates in wing folding. Pluses and minuses in parentheses without furrow names indicate convex and concave secondary furrows participating in wing folding. nl, nodal line. Scale 2 mm.

or even flex their wings, for example, the dipteran louse fly *Lipoptena cervi* L. (Muscida: Hippoboscidae) and soldier fly *Pachygaster atra* Panzer (Stratiomyidae) (fig. 4.3*E, I*), and the hymenopteran *Acantholyda nemoralis* Thomson (Pamphilidae) and *Urocerus gigas* L. (Siricidae) (fig. 4.4*B*).

Furrow morphology has never been discussed in detail. Martynov (1924) defined a furrow as a "narrow band of refined chitin." A similar viewpoint was expressed by Rasnitsyn (1969). However, electron and light microscopic studies show that this is true only for furrows on secondarily thickened wings, which serve a defensive function. In other cases, the site of wing bending is as thin as the surrounding membrane (Grodnitsky 1991a). Usually a furrow can be seen by examining the wing in transmitted light, because of the corrugation of cuticular membrane in the zone of furrow location. In general, furrow morphology proves to be quite diverse (it is discussed in sec. 4.7, after a presentation of relevant empirical material).

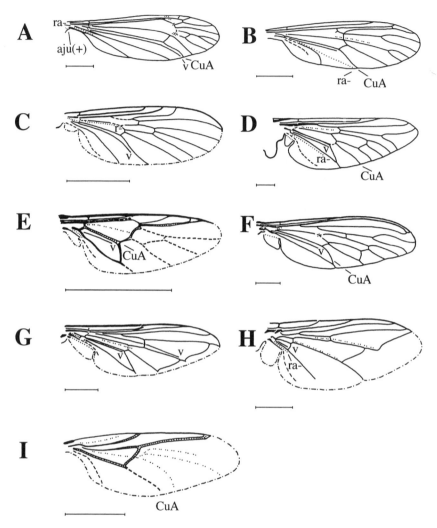

Figure 4.3. Wing morphology of dipterous flies. NEMATOCERA: (*A*) *Nephrotoma* sp. (Tipulidae); (*B*) *P. lacustris* Mg. (Ptychopteridae); (*C*) *Anisopus* sp. (Anisopodidae). BRACHYCERA ORTHORRHAPHA: (*D*) *H. pavlovskii* Ols. (Tabanidae); (*E*) *P. atra* Panzer (Stratiomyidae); (*F*) *Asilus* sp. (Asilidae). BRACHYCERA CYCLORRHAPHA: (*G*) *I. glaucius* L. (Syrphidae); (*H*) *T. magnicornis* Zett. (Tachinidae); (*I*) *L. cervi* L. (Hipposcidae). Doubled dotted line, nonsclerotized membrane gutter; v (vena), secondary false vein. Corrugation on veins is shown with transverse dashes. Other designations as in figure 4.2. Scale 2 mm.

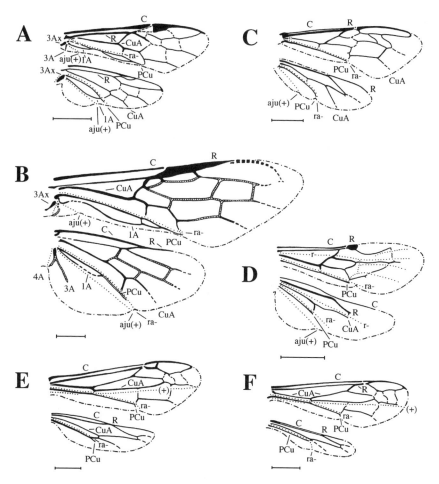

Figure 4.4. Wing morphology of hymenopterans: (*A*) *Rhogogaster* sp. (Tenthredinidae); (*B*) *U. gigas* L. (Siricidae); (*C*) *Apis mellifera* L. (Apidae); (*D*) *Tiphia femorata* F. (Tiphiidae); (*E*) *Ancistrocerus parietinus* L. (Eumenidae); (*F*) *Dolichovespula silvestris* Scop. (Vespidae). 3Ax, 3rd axillary sclerite. Other designations as in figures 4.2 and 4.3. Scale 2 mm.

4.2.   Wing Morphology and Evolution in the Amphiesmenoptera

Wing venation and topology of furrows are most diverse in the Amphiesmenoptera—a taxon that includes two closely related orders: the caddisflies Phryganeida, and the moths and butterflies Papilionida.

4.2.1. *Nomenclature of Veins and Furrows*

Among all existing nomenclatures of longitudinal veins, the version most widely used at present is that developed by Comstock and Needham (1898, 1899; Comstock 1918), and modified by Snodgrass (1935). Hamilton's (1972a, 1972b) nomenclature, sometimes used for entomologic education and incorporated in certain textbooks (Ross et al. 1982), received deserved criticism from Emelianov (1977) and cannot be accepted. The proposals of Hamilton are fundamentally unsupported by argumentation. His system, in general, suggests a noncritical hybridization of the nomenclatures of Forbes (1943) and Snodgrass (1935). Hereinafter, the designations of Comstock, Needham, and Snodgrass are used.

The wings of primitive caddisflies and moths bear eight primary longitudinal veins: costa (C); Sc; R with branching RS; M with two branches, media anterior (MA) and media posterior (MP); Cu with CuA and CuP; postcubitus (PCu); and two anal veins, 1A and 2A (fig. 4.5). Hamilton postulated independence of the RS and CuP, giving them the status of primary veins. This viewpoint contradicts data on vein paleontology (Martynova 1960) and comparative morphology, including tracheation of the veins (Emelianov 1977). The cubital branches are indeed quite different in appearance. The CuA is generally two-branched, well developed, and sclerotized. Together with the Sc and R, it constitutes the main body of the remigium. In cases in which the CuA is weakened, it is mechanically enhanced by means of precubital thickening. In contrast to the CuA, the posterocubital vein is always considerably weakened and thin, and contains little melanin pigment. The concave furrow passes along its entire length and separates the remigium from the clavus. Consequently, the anterior and posterior cubital veins are functionally different: CuA reinforces the wing, whereas CuP is initially associated with the longitudinal furrow, cannot enhance wing rigidity, and is reduced in the majority of lepidopterans with advanced venation.

In many caddisflies (Ivanov 1985b, 1987a, 1987b) and in some primitive moths (Hepialidae: Sharplin 1963a, 1963b), the functional difference of the CuA and CuP results in their departing independently from the axillary sclerites instead of merging basally. If this were their primary relationship, Hamilton's (1972a) viewpoint could be accepted; following Forbes (1943), he designated the CuP as an independent plical vein. However, CuA and CuP are fused proximally in the supposed ancestors of caddisflies, from Paoliida (fig. 1.2) to Mecoptera (fig. 4.6*B*). Hence, the independence of cubital bases

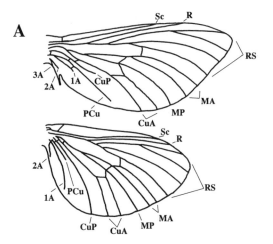

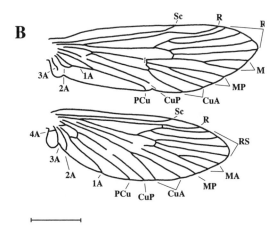

Figure 4.5. Wing venation of primitive moth and caddisfly: (A) *Pielus labirinthecus* (Hepialidae) (from Comstock 1918); (B) *R. nubila* Zett. (Rhyacophilidae). Designations as in figure 4.2. Scale 2 mm.

likewise originated secondarily in response to the functional differentiation between the CuA and CuP.

The postcubital and anal veins are situated behind the Cu. The independence of the PCu in relation to 1A and 2A is well substantiated and unquestionable: in the generalized state, this vein always comes from the third axillary sclerite (3Ax), irrespective of the anal veins (Emelianov 1977; Ivanov 1994). In this instance, Hamilton's proposal could be accepted, and the PCu could be termed the *empusal vein*, since the PCu has no relation to the Cu, located in another functional zone of the wing (clavus) and being related to another part of axillary appartus, namely 3Ax, but not with medial plates, as cubital veins. In any case, this name change would only have a purely seman-

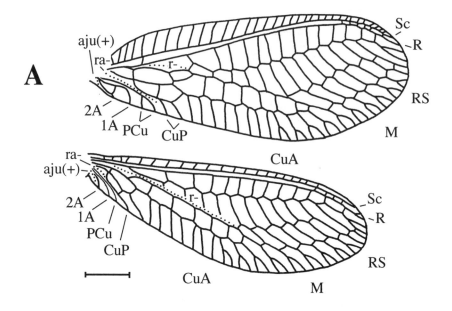

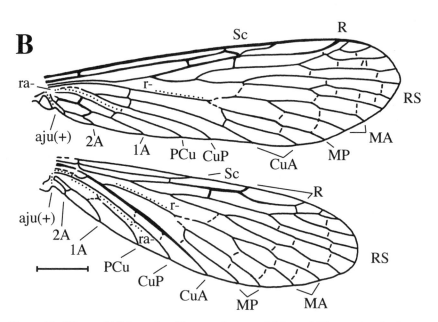

Figure 4.6. Wings of (*A*) lacewing *Chrysopa dasyptera* McL. and (*B*) scorpion fly *P. communis* L. Designations as in figure 4.2. Scale 2 mm.

tic value. The number of anal veins in lepidopterans and caddisflies varies, but is initially equal to two. In the majority of cases neither the PCu nor 1A or 2A bifurcate.

In the nomenclature proposed by Emelianov (see also Brodsky [1987, 1988]), furrows are named according to the veins that pass parallel in front of them. A basic disadvantage of this system of designations is that the same furrow can be situated differently relative to a given vein. A furrow can cross an accompanying vein and then travel either behind or in front of it (fig. 4.7A–C). A furrow can be expressed in some species as a band of light soft membrane passing at both sides of a vein (fig. 4.7A, B, D). Furthermore, the veins that are associated with furrows are often reduced. For example, in lepidopterans the partial or complete reduction of M and CuP is quite common (fig. 4.8). In these cases the direct use of the Emelianov nomenclature principle leads to situations in which the same furrow acquires different names; the medial furrow becomes the furrow of RS, whereas the CuP becomes an anterocubital furrow.

Wootton (1979) suggests inadequate terms. For example, it is unclear what distinguishes his "line" and his "furrow." Furthermore, he classifies furrows according to their functions, differentiating "flexion lines" and "fold lines." Such a functional approach to classifying morphologic structures is inappropriate from my viewpoint, because flexion lines too often secondarily participate in wing folding, as will be shown later. Consequently, I will use the nomenclature of Martynov (1924), in which a furrow is designated according to the wing zone within which it is located, or according to the zones that are demarcated by the furrow.

In Martynov's nomenclature, the three primary furrows are termed *remigial* (plica remigialis), *remigio-anal* (p. remigio-analis) (concave), and *ano-jugal* (p. ano-jugalis) (convex) (fig. 4.7). The concave furrows are adapted for flapping flight and help form an aerodynamically advantageous wing profile during upstroke (sec. 2.1). The convex furrow is necessary to fold up the wings when at rest.

4.2.2. Transformation of Wing Planform

The initial shape of caddisfly wings probably differed somewhat from the wing shape of recent generalized species. Thus, many fossil Phryganeida had wings with a round tip (Sukacheva 1982), exactly like recent Panorpida. This suggests that the common ancestors of caddisflies and lepidopterans

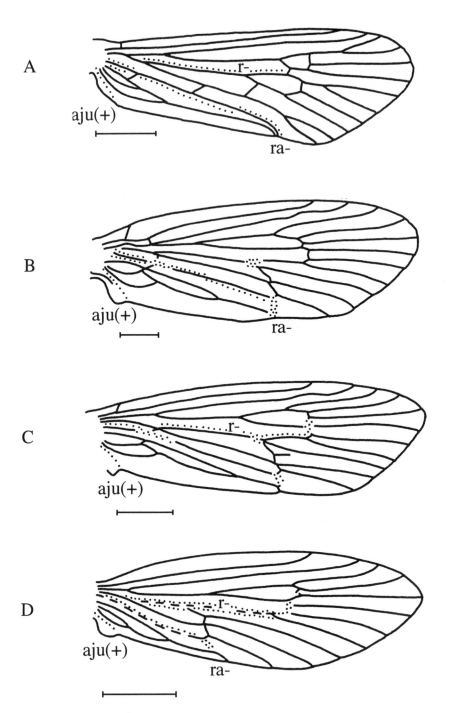

Figure 4.7. Forewings of caddisflies: (*A*) *H. nevae* Kol. (Hydropsychidae); (*B*) *Phryganea bipunctata* Retz. (Phryganeidae); (*C*) *Agrypnia pagetana* Curt. (Phryganeidae); (*D*) *Lepidostoma hirtum* F. (Lepidostomatidae). Designations as in figure 4.2. Scale 2 mm.

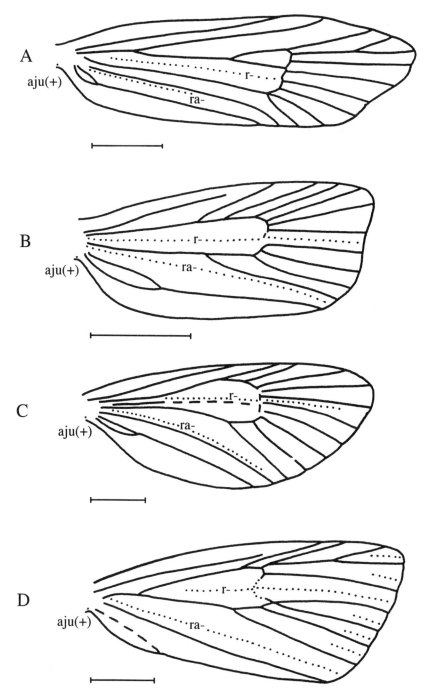

Figure 4.8. Forewings of moths: (*A*) *Nemapogon dorsiguttella* Ersch. (Adelidae); (*B*) *Laspeyresia pomonella* L. (Tortricidae); (*C*) *I. sinensis* Walk. (Zygaenidae); (*D*) *Agriphila straminella* Den. et Schiff. (Crambidae). Designations as in figure 4.2. Scale 2 mm.

had wings with pronounced humeral and jugal, and smooth apical and anal angles.

According to available data, moth wings initially had a planform typical of many primitive caddisflies. The same wing configuration characterizes small ghost moths (e.g., *Alphus sylvinus* L., Hepialidae). Large ghost moths also have distinct anal and jugal angles, although their wings are secondarily widened. The planform of the wings of primitive moths, suggested as symplesiomorphic for the order (Kozlov 1987, 1988), should be considered to be a specialization originating with the decrease in body size, because the wings of many tiny species are transformed in a similar way (fig. 4.9 *A–C*). The presence of secondary features in the organization of the wings of primitive moths is also indicated by the partial loss of their venation (reduction one of the anal veins, as compared with ghost moths and other primitive lepidopterans) considered subsequently.

Apart from the width of the wing hinge, other parameters of wing shape lack even a hypothetical functional interpretation (Grodnitsky 1995b). As already mentioned, species of many moth families secondarily evolve broad wings. Their origin is explained variously: in diurnal ectotherms by the necessity to accumulate heat from the sun, and in nocturnal endotherms by the tendency to decrease the energy expenditures of a nonfeeding imago with reduced mouthparts (sec. 3.8). Nevertheless, comparison of diurnal and nocturnal broad-winged lepidopterans from different families reveals significant similarity of the wings' planform, including secondary marginal projections (fig. 4.10). Analogous examples can be presented on the lanceolate wing shape, typical of tiny caddisflies and moths (fig. 4.9*A–C*) and unusual in other orders, but characteristic of the moth flies Psychodidae (fig. 4.9*H*) and the spear-winged flies Lonchopteridae. These convergent features, however, are presently unexplainable from the perspective of functional similarity.

4.2.3. *Transformation of Venation*

Initially caddisflies and moths were insects of medium size, with wingspans of 10–15 mm (Sukacheva 1982; Kozlov 1988). During the course of evolution, the body size of different groups increased or decreased, which considerably influenced the historical trajectory of wing venation. The most abundant venation characterizes primitive species of medium size; the number of veins diminishes in advanced groups of both large and small size (Grodnitsky 1991a).

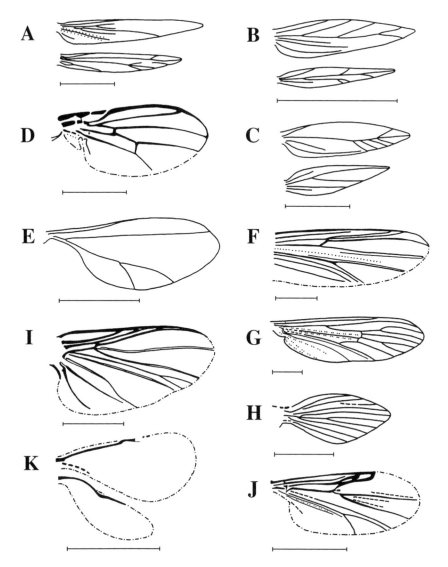

Figure 4.9. Wings of small insects: (*A*) caddisfly *Hydroptila vectis* Curt. (Hydroptilidae); moths (*B*) *Stigmella kozlovi* Pupl. (Nepticulidae), and (*C*) *Bohemannia ussuriella* Pupl. (Nepticulidae); dipterous flies (*D*) *Drosophila* sp. (Drosophilidae); (*E*) *Dasyneura laricis* Rozhkov (Cecidomyidae); (*F*) Chironomidae Gen. sp.; (*G*) *Aedes cyprius* Ludl. (Culicidae); (*H*) *Psychoda* sp. (Psychodidae); (*I*) *Odagmia ornata* Mg. (Simuliidae); (*J*) *Culicoides grisescens* Edw. (Ceratopogonidae); wasp (*K*) *Discodes aeneus* Dalm. (Encyrtidae). Marginal fringe is not shown. Scale 1 mm.

Figure 4.10. Planform of lepidopteran wing couple: (*A*) lemon-yellow butterfly *Gonepteryx rhamni* L. (Pieridae); (*B*) measuring worm moth *Metrocampa margaritata* L. (Geometridae). (From Lampert 1913.)

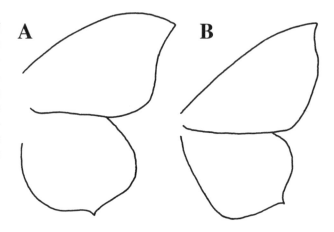

Vein reduction is most clearly observed under a considerable decrease in wing size (Gorodkov 1984) (fig. 4.9). For example, the longitudinal veins of both caddisflies and moths partially disappear after the wing becomes shorter than 3 to 4 mm (Grodnitsky 1991a), a threshold value below which the inherent mechanical properties of the cuticle are sufficient to provide the necessary wing strength with partial or even complete vein reduction. Vein reduction can also be explained by a general embryologic regularity: the decrease in absolute size of a vestige below a particular limit prevents differentiation of the vestige (Lande 1978; Alberch 1980; Alberch and Gale 1985; Beloussov 1987; Oster et al. 1988). Cross sections of insect wings reveal that a vein that is being reduced gradually loses its initially tubular structure and becomes flatter, its walls get thinner, and sclerotization decreases. Finally, only a convex (or concave) gutter-like relief structure remains in the adjacent membrane, after which the vein disappears completely. Thus, the reduced wing venation of tiny insects offers an illustration of the collaborative action of additive morphogenetic and selective factors of organismal appearance. Various factors can influence a decrease in body size, for example, transition to life in a limited space (inside plant leaves, insect eggs, etc.), and diminution in turn directly causes vein loss. If the small wing has sufficient strength to fly without veins, miniature species evolve with reduced venation. If the strength had been insufficient, such species would either have originated with novel special mechanisms of vein differentiation under small size or not originated at all.

Miniaturized forms repeatedly originated in different phyletic branches of the Amphiesmenoptera. In each case, reduction of different veins took place, so that their strict homologization is possible only by examining a series of closely related species that demonstrate a gradual size decrease. The majority of tiny species, as a rule, have different combinations of Sc, R with one to four branches, Cu, PCu, 1A, and 2A on their forewings (fig. 4.9A–C).

Caddisflies generally have much more generalized venation. Among species of normal size, some Leptoceridae (*Triaenodes bicolor* Curt.) and Molannidae (*Molanna angustata* Curt., *Molanna albicans* Zett.) can lack one to three branches of RS on their forewings, whereas the initial number of RS branches is four. The most frequently reduced veins on trichopteran hindwings are one to three branches of RS, one to two branches of M, and one of the anal veins.

Flight in large lepidopterans is more highly developed than in caddisflies. Correspondingly, their wings display a greater degree of transformation with any given increase in insect size. The forewings of moths and butterflies have often lost one to two branches of RS, one branch of M, CuP, and 1A and 2A, whereas the hindwings often lack one to four branches of RS, one to two branches of M, CuP, PCu, and both 1A and 2A. In both moths and caddisflies, the hindwings have become noticeably stronger than the forewings. This change reflects the presence of wing coupling mechanisms and, consequently, the loss of active flight function by the hindwing and the transfer of the main functional activity to the mesothorax and forewing.

The reduction of the posterocubital vein is extremely characteristic of moths (Grodnitsky 1991a). In generalized species, it is distinct and associated with the adjacent furrow. This proximity probably makes the vein irrelevant as a strengthening element and induces its reduction on both fore- and hindwings in even primitive forms. This reduction occurs in the proximal part of the medial vein as well.

Reduction of the first two anal veins on the forewings often passes through an intermediate stage in which they fuse and flow into the PCu instead of the trailing edge of the wing, so that the so-called "anal loop" is formed (figs. 4.7 and 4.8). The three fused veins are always situated within the clavus—the zone separated from the rest of the wing by the concave remigio-anal furrow in the front and by the convex ano-jugal furrow behind. The anal loop is observed in the majority of caddisflies and moths from the most primitive families: Eriocraniidae (Davis 1978; Davis and Faeth 1986), Heterobathmiidae (Kristensen and Nielsen 1979), and Agathiphagidae (Kristensen 1981a).

Brodsky and Ivanov (1983b) suggested that the anal loop originated owing to fore- and hindwing coupling. The evolution of rhynchote insects illustrates this thesis well: in the three phyletic lines of the order (cicadas + bugs, jumping plant lice + whiteflies, and aphids + coccids), the loop was formed independently and simultaneously with the appearance of coupling mechanisms (D. E. Shcherbakov, personal communication). Another assumption exists: in caddisflies the anal veins lose connection with the C so that a continuous covering is formed by the left and right wings at rest. As a result, air can be retained under the wings and the imago obtains a novel adaptation for life underwater (Sukacheva 1982).

None of these explanations are adequate. A well-developed loop is present on the wings of primitive moths (a suggested apomorphy of Papilionida; see Kristensen [1984]), although the wings of these insects are not coupled during flight (see chap. 1 and sec. 3.8), and their imagos are not associated with aquatic habitats. A similar structure was found on the forewings of some walking sticks (Ragge 1955), which are functionally four-winged primarily terrestrial insects. The loop also occurs on the hindwings of fossil and recent caddisflies (Novokshonov 1992), recent lepidopterans (Grodnitsky 1991a), and hymenopterans (fig. 4.4*A, C*). This fact lacks even a preliminary adaptive explanation. The previous considerations cannot be allowed as refutation of the existing interpretations of anal loop function, because there is no evidence to deny that in one taxon it serves as strengthening of the wing, whereas in another it constitutes an adaptation for an amphibiotic mode of life. However, these functional interpretations can only explain the conservation of a loop that had already emerged. The question of the origination of novel features is of much greater importance to evolutionary theory than explanations of their subsequent maintenance (Maynard Smith 1978; Lubischew 1982).

A sclerotization is located in the jugal regions of the wings of primitive species. Martynov (1924b, 1925) showed that the hindwing jugum plays an important role in the evolution of the flapping surface of insects, while the jugal elements (vena cardinalis and v. arcuata according to Martynov) develop into true tubular veins that reach the wing edge (fig. 4.1). The existence of expressed jugal venation was also recognized by later investigators (Snodgrass 1935; Shwanvich 1949; Sharov 1968). However, the independence of the jugal relative to the anal veins has not been conclusively proved. As a rule, both the anal and the jugal veins depart from a single basement, separated by the ano-jugal furrow in the majority of species into two interrelated parts (this is a usual argument for contrasting anal venation to jugal). Meanwhile, in some caddisflies (e.g., in *Brachycentrus subnubilus* Curt. [Brachycentridae] and *Ganonema extensum* Mart. [Calamoceratidae]), the furrow passes in front of the hinge and does not divide it. Thus, it is more reasonable to agree with Rasnitsyn (1969) that the jugal veins result from secondary differentiation of the anal system, and designate v. cardinalis as 3A (fig. 4.5). Furthermore, there is no reason to deprive v. arcuata (the axillary cord according to Rasnitsyn) of the status of an independent vein: in lepidopterans it has been lost, but on the hindwings of many caddisflies it is sclerotized, quite well developed, and can be termed 4A (fig. 4.5B).

The jugal region of the Phryganeida and the Papilionida occasionally underwent evolutionary changes related to specialization for different functions. Comstock (1918) thought that the elongated jugums of ghost moth forewings served as coupling with the hindwings. Martynov (1924b) disagreed, claiming that jugums were structurally too weak. It is now clear that Martynov was correct, because the wings of primitive moths are not coupled and work separately in-flight (sec. 1.2 and 3.8). Consequently, the jugum does not act as a part of a coupling mechanism. (The jugal region of the forewings is lost in some caddisflies and in the majority of lepidopterans.)

Evolution of jugums on the hindwings was somewhat different. Cinefilming and photographic surveys of the flight of lepidopterans and caddisflies show that the hindwing jugums are almost motionless relative to the body during the course of a stroke cycle. Basically strokes are executed about the axis passing through the axillary apparatus and ano-jugal furrow. Therefore, the hindwing jugum is in fact outside of the flapping surface and does not directly generate aerodynamic force; the jugum probably blocks airflow between the flapping surface and the abdomen. Owing to differences of pressure over and under the body, such a flow would originate in the absence of a jugum and reduce lift.

A broadened ano-jugal zone is used in an alternative manner by many butterflies of the superfamily Papilionoidea. Their jugum bears a specialized, two-layer scale covering that consists, in part, of shiny light-reflecting scales. The jugums adhere to the abdomen at rest and to some degree during flight, thereby functioning as insulators to prevent heat loss from the insect's body.

A single vein is located on the rear part of the hindwings of many moths and butterflies in the zone separated by the anterior convex ano-jugal furrow (fig. 4.11A, C–E). This is probably 3A, which is always situated behind the convex fold line. Generally the third anal vein persists on the hindwing even after the reduction of 1A and 2A.

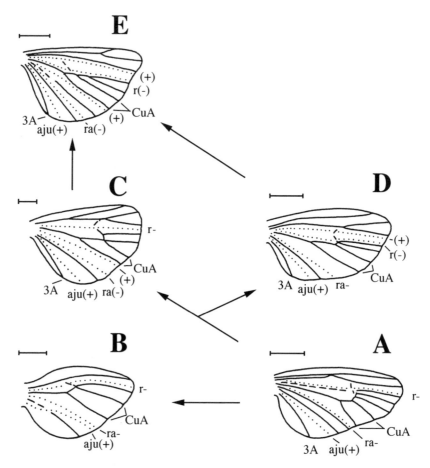

Figure 4.11. Evolutionary transformation of hindwing folding system in moths: (A) *I. sinensis* Walk. (Zygaenidae); (B) *S. phygea* L. (Syntomidae); (C) *Phragmatobia fuliginosa* L. (Arctiidae); (D) *Archips podana* Scop. (Tortricidae); (E) *Margaritia sticticalis* L. (Pyraustidae). Designations as in figure 4.2. Scale 2 mm.

4.2.4. Transformation of Furrow System

All three primary furrows are usually clearly visible in primitive recent caddisflies and moths. The only exceptions are the smallest species, which often lose the plica radialis. Also distinct are the thyridium and arculus, which are desclerotized spots or fenestrae located accordingly on the middle of the M and in the end of the CuP (fig. 4.1B, C; fig. 4.7B, C). An invisible nodal line passes through the fenestrae, along which the wing tip bends ventrally at the end of downstroke. Another fenestra is often situated at the base of the terminal branches of the RS or M. It participates in the formation of the trapezia-like deformation of the wing tip (fig. 4.1B; fig. 4.7C; Kristensen and Nielsen 1979; Grodnitsky and Kozlov 1985; Davis and Faeth 1986).

The primitive state of the furrows also changed in respect to the different evolutionary transformations of the wings observed in the Amphiesmenoptera (sec. 3.8). The furrows are often reduced in tiny species (fig. 4.9A–C). This reduction can be related to the feather-like wings that often originate with a decrease in absolute size (sec. 3.8): the wing becomes narrow, and the flapping surface consists mostly of a long fringe, which works as a solid wing.

Ventral flexion and trapezoid deformation of the wing tip have been observed in the majority of investigated caddisflies and moths, although often independently of visible morphologic features. Thus, the plume moths *Capperia* sp. and *Oidaematophorus* sp. (Pterophoridae) possess a fundamentally identical flapping surface morphology. However, the first species exhibits sharp ventral bending, whereas the latter displays a smooth flexion (Grodnitsky and Morozov 1994). Remarkably the fenestrae (thyridium and arculus) are not expressed in either species.

Because of different ecologic reasons (sec. 3.8), an increase in body size is followed by a widening of the flapping surface in many groups of the Amphiesmenoptera. In the majority of cases, this results in secondary furrows, some of which participate in compact plaiting of the wings at rest. These furrows pass from wing hinge to edge, and are called *general* in Emelianov's (1977) terminology. Primitive concave furrows often participate in hindwing plaiting. These include p. remigialis in Tortricidae (Papilionida); p. remigio-analis in Arctiidae (Papilionida), Phryganeidae, and some Calamoceratidae (Phryganeida); and p. remigialis and p. remigio-analis in Pyralidae, Crambidae, Phycitidae, and other pyraloid lepidopterans (figs. 4.11 and 4.12).

Other furrows are situated wholly in the distal third of wings parallel to the terminal vein branches (fig. 4.8D). Brackenbury (1991a, 1994) states that they serve for wing deformation and designates them flexion lines. Meanwhile, his published photographs of free-flying butterflies obviously demonstrate that the profile change does not occur along the furrows, but rather by means of flexion of the membrane that surrounds the furrows. The peripheral furrows are not associated with any change in the flapping surface profile. These furrows are apparently necessary to enhance wing strength, a property that separates them from fold lines and flexion lines of different types.

Secondary general furrows originate even in species in which body size increase is unaccompanied by any broadening of the wing surface. For ex-

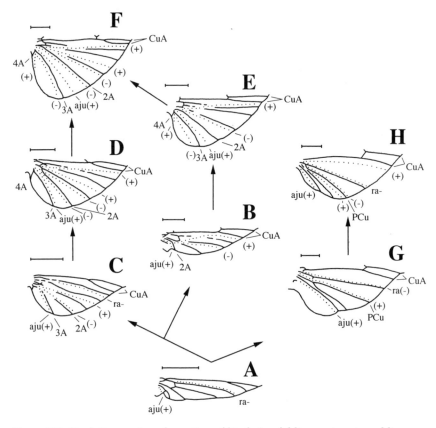

Figure 4.12. Evolutionary transformation of hindwing folding system in caddis-flies; remigium is not shown. (A) *R. tristis* Pict. (Rhyacophilidae); (B) *R. vulgaris* Pict. (Rhyacophilidae); (C) *S. nigricornis* Pict. (Goeridae); (D) *C. villosa* F. (Limnephilidae); (E) *A. laevis* Zett. (Limnephilidae); (F) *Glyphotaelius pellucidus* Retz. (Limnephilidae); (G) *P. bipunctata* Retz. (Phryganeidae); (H) *G. extensum* Mart. (Calamoceratidae). Designations as in figure 4.2. Scale 2 mm.

ample, the primitive caddisflies *Rhyacophila nubila* Zett. and *Rhyacophila tristis* Pict. have only three primary furrows (fig. 4.12A). *Rhyacophila vulgaris* Pict. is larger but has a virtually identical flapping-surface morphology. However, its hindwing bears two additional furrows along which the wing folds up at rest (fig. 4.12B).

A widening of the flapping surface is very common in the Amphiesmenop-tera and occurs in two alternative ways in caddisflies and lepidopterans. In the Phryganeida, only the ano-jugal field of the hindwings, or vannus, is broadened, whereas the entire forewing and the remigium of the hindwing remain relatively unchanged. By contrast, in the Papilionida both the fore- and the hindwing shape is changed. In the majority of cases, the widening of a flapping surface requires that it be packed compactly while inoperative, resulting in the appearance of new general furrows on the hindwings. Different methods of increasing wing area resulted in distinctive systems of wing plaiting in the Phryganeida and Papilionida; therefore, the evolution of the furrow systems of caddisflies and moths will be considered in turn.

4.2.4.1. Caddisflies

The wings of generalized caddisflies fold up only along one convex ano-jugal furrow (fig. 4.12A). Increase in body size, and the appearance and further broadening of the hindwing vannuses, led to the development of two secondary fold lines, convex and concave (fig. 4.12B, C). Their location relative to the main longitudinal veins can vary in different species. For example, it is possible to trace how a convex furrow first crosses the PCu, then the CuP and gradually approaches the CuA in the series of intermediate forms *Apatania wallengreni* McL. (Limnephilidae), *Silo nigricornis* Pict. (Goeridae), *Brachycentrus subnubilus* Curt. (Brachycentridae), and *R. vulgaris* Pict. (Rhyacophilidae). At the intersections, the veins are soft and lack melanin pigmentation. No cases in which the furrow crosses the CuA are observed. Functional constraints presumably prohibit any weakening of this vein.

In the majority of caddisfly species, the convex anal furrow crosses the concave primary p. remigio-analis furrow as it shifts toward the Cu. This crossing results in the formation of a proximal deltoid zone on the wing, separated by the two furrows from the basal portion of the wing. Presumably communication of control forces from the axillary sclerites to this zone is handicapped, and problematic deformation of the wing can arise during plaiting or flying. Furrow crossing can be avoided by eliminating one of the furrows. This mechanism probably caused the reduction of the remigio-anal furrow, which has disappeared in many species (fig. 4.12B, D–F). A unique way to avoid crossing of the furrows was observed in *Ceraclea nigronervosa* Retz. (Leptoceridae). In this species, p. remigio-analis is pressed closely to the anterocubital vein and located between the CuA and the secondary convex claval furrow.

Further widening of the ano-jugal region results in the development of two more pairs of furrows. In some caddisflies, convex and concave furrows are acquired by an enlarged clavus (fig. 4.12D). In Phryganeida with a broader jugum, two new jugal furrows (one convex and one concave) appear (fig. 4.12E). Finally, in caddisflies with the widest vannus, the hindwing at rest folds up along all seven of the described furrows (fig. 4.12F).

All the wing plaiting systems modifications described in the foregoing have a common basis, because the control of hindwing folding is carried out by the upward turn of the distal arm of 3Ax. The mechanical effort is transmitted from the sclerite to the base of 2A, which has a bend in its proximal part; all furrows that participate in wing folding begin from this bend (fig. 4.12B, C, E, F). The plaiting of the flapping surface is achieved in a different manner in the family Phryganeidae, as well as in other species. In those caddisflies, the folding effort is transmitted from 3Ax through the base of the PCu, rather than via 2A (fig. 4.12G). In contrast to the first case, this folding system incorporates the concave p. remigio-analis furrow, which secondarily participates in wing plaiting. Further development of a flapping surface of the phryganeid type leads to the formation of one more pair of furrows, convex and concave (fig. 4.12H).

Thus, two systems of plaiting of the ano-jugal zone are found in caddisflies, differing by the point of application control forces in the proximal part of the wing. Different location of furrows affects the venation of the basal area. In the first case, the proximal parts of the PCu and 1A are often reduced, whereas formation of the second type of folding system can be followed by loss of the 1A and 2A bases. Nevertheless, in both cases formation of a morphologically similar structure takes place, namely the vein proximal bend that transmits force from 3Ax to the furrows.

4.2.4.2. Moths

The hindwings at rest fold up only along the ano-jugal furrow in generalized moths (fig. 4.11A), exactly as they do in caddisflies. A modification of this plaiting system is found in the wasp moth *Syntomis phygea* L. (Syntomidae), whose hindwings bear p. ano-jugalis beginning behind the anal vein base, then declining forward and crossing 1A, coming onto the wing margin just behind p. remigio-analis (fig. 4.11B). Consequently, wasp moths possess a plaiting system similar to that of many caddisflies.

A broadening of the moth flapping surface results first in the appearance of the secondary claval convex general furrow, while the wings acquire the ability to bend also along the primary concave p. remigio-analis, that is, along the furrow that initially performed a strictly aerodynamic function without participating in plaiting (fig. 4.11C). Broadening of the remigium is accompanied by the formation of an additive convex remigial furrow that passes anterior to the primary concave p. remigialis (fig. 4.11D). Finally, up to five furrows occur in species with the most well-developed wing folding system (fig. 4.11E). These furrows, sometimes in various combinations, are characteristic of many of the Tortricidae, Crambidae, Pyralidae, Choreutidae, Limacodidae, Noctuidae, Arctiidae, Lymantriidae, and other moths that fold up their hindwings at rest. Thus, the methods of wing plaiting differ considerably in caddisflies and moths, reflecting each order's characteristic manner of broadening the flapping surface.

4.3. Wing Morphology in Neuropterans, Scorpion Flies, and the Hypothetical Ancestor of the Amphiesmenoptera

The functional organization of the wings of the Myrmeleontida and Panorpida is quite similar and corresponds to the most primitive anteromotoric functionally four-winged kinematics. An interesting problem is the comparison of their wings with the morphology of a hypothetical amphiesmenopteran ancestor reconstructed in agreement with the data on recent species.

The wings of scorpion flies and neuropterans are connected to the body with a narrow hinge. In many forms the hinges are further narrowed, so that the jugal angle of the wings completely disappears. This feature is not specific to caddisflies and lepidopterans but can be considered a homoplasy, evolving in many different insect groups.

The primitive state of amphiesmenopteran venation is, in all likelihood, similar to the construction of the wings of fossil caddisflies (Sukacheva 1982) and recent scorpion flies. They are characterized by comblike Sc, R with four-branched sector, M with two major branches, each also divided into two veins, fore and hind cubituses, PCu, and two anal veins. Many veins, especially the remigial, may bear additional terminal bifurcations.

The wings of Papilionida may be completely derived from the construction of wings of recent primitive Phyganeida and differ in general only by the absence of M4 on forewings—primitive for moths (Kozlov 1988). The presence of M4 in lepidopterans belonging to the family Agathiphagidae (Kristensen 1981a) is probably a secondary feature, repeating the ancestral condition.

Furrows present on the wings of generalized caddisflies and moths reflect the initial construction of the wings of the amphiesmenopteran ancestor, and their presence may be primitive for all winged insects. The wings of the Pterygota probably acquired the ability to fold up before the evolution of active flapping flight (sec. 3.1). Simultaneously, the ano-jugal furrow had to originate. The appearance of two concave primary furrows (remigial and remigio-anal) could occur somewhat later, in close relation with the formation of flapping flight. Thus, the wings of ancestors of the Pterygota needed to bear three lines of deformation: p. remigialis, p. remigio-analis, and p. ano-jugalis. This hypothesis is partly supported by the configuration of the marginal edge of the wings of early fossil insects. In fact, in recent species the location of furrow and costal vein connection is often marked by an incision (fig. 4.1*A, C;* fig. 4.8*A, B;* fig. 4.11*A, D, E;* fig. 4.12*E, G*). This presumably facilitates wing deformation along the furrows. Similar incisions are located very close to the end of the CuP on the wings of carboniferous Paoliida (Rasnitsyn 1980), providing evidence for the presence of the concave remigio-anal furrow on the wing and, hence, of flapping flight in paoliid insects.

In the Myrmeleontida and Panorpida, the remigial and remigio-anal furrows of the fore- and hindwings are poorly developed and can only be seen close to the hinge, being absolutely invisible in their distal parts (fig. 4.6). Remigial furrows in *Panorpa communis* L. are almost imperceptible. Bifurcation of the medial vein in the scorpion fly is represented by considerably weakened parts of veins and is analogous to the thyridium (the fenestrum on caddisfly and lepidopteran wings). The cross veins of the distal half of the wings are also weakened (fig. 4.6*B*). This can favor the formation of a gradual concave profile observed in these insects during their second half-cycle (Grodnitsky and Morozov 1994). In general, the neuropterans and scorpion flies possess features of primitive organization, and a quite strong similarity can be seen between them and the hypothetical ancestor of the endopterygote insects (fig. 1.6). In addition, lacewings and especially scorpion flies are characterized by a set of apparently secondary features that determine the relatively weak deformation of wings that is most likely functionally connected with the apomorphic narrowing of the wing hinges.

4.4. Wing Morphology in Dipterous Flies

Muscida wings can be considered as monofunctional structures, since they are not used in thermoregulation, for defensive purposes or for other flight modes besides flapping. Hence, the wing shape in dipterous flies must be essentially close to the optimal condition defined by the function of flapping flight. Accordingly, the planform of dipteran wings is considerably more uniform throughout the order than in the related groups—the orders Phryga-neida and Papilionida.

Vein homology of dipteran wings will be discussed using the system pro-posed by Wootton and Ennos (1989). To homologize venation elements, they used the arrangement of furrows and showed that the vein located behind the CuA and generally designated as CuP is, in fact, a false vein, represented by a sclerotized gutter rather than a tube. Let us consider the problem of the origination of false veins.

Muscida wings often bear secondary relief gutter-like structures such as the precubital thickening and peripheral furrows of caddisflies, moths, and butterflies. It is quite common in entomology to associate formation of such structures with the necessity of wing strengthening (Ussatchov 1970; Woot-ton 1981; Brodsky and Ivanov 1983b; Kukalova-Peck and Lawrence 1993). This viewpoint is natural in respect to dipterous flies, with their high stroke frequency. Gutters formed by the membrane are situated before the M in *Ptychoptera lacustris* Mg. (Ptychopteridae) (fig. 4.3B), in *Hybomitra pavlovskii* Ols. (Tabanidae) (fig. 4.3D) and in many other horse and deer flies, after the M in *Ischirosyrphus glaucius* L. (Syrphidae) (fig. 4.3G), before the M and before the Cu in *Tachina magnicornis* Zett. (Tachinidae) (fig. 4.3H), and in the anal zone in mosquitoes of the genus *Aedes* (fig. 4.9G). In many cases, these gut-ters are strengthened through additive thickening and sclerotization of the membrane, thereby acquiring similarity with veins. A well-known example is the extra vein, vena spuria, most distinctly expressed in flower flies Syr-phidae (fig. 4.3G) and in some phantom crane flies Ptychopteridae. Notably, sclerotization of the v. spuria can vary considerably between different flower flies. It is strong in *Chrysotoxum, Eristalis, Syrphus* (sensu stricto), *Eriozona,* and *Posthosyrphus* and much weaker in *Sphegionoides, Temnostoma, Ischirosyr-phus, Epistrophe,* and *Helophilus.* This interspecific variability of the false vein probably reflects different degrees of secondary modification of an initially unsclerotized gutter. Extra "veins" never have tubular structure; they can be either convex (as v. spuria) or concave (as the false vein locating after the Cu).

Although it seems logical to explain the appearance of additive relief as a mechanism to increase wing strength, this analysis encounters contradic-tions. Structures similar to false veins are also present in small—and even in the smallest—forms, whose wings are sufficiently strong for flapping flight without venation. As already noted, lepidopterans and caddisflies begin to lose veins when the wing length decreases below 3 to 4 mm. The greatest degree of reduction typifies tiny wasps (fig. 4.9K), thrips, many rhynchote insects, and some dipterous flies (fig. 4.3I; fig. 4.9E).

However, the venation of dipterous flies is, in general, quite conservative and is not necessarily reduced at miniaturization. In some nematoceran dipterous flies, the wings bear additional corrugation, which sometimes obtains additional sclerotization. In biting midges Ceratopogonidae (fig. 4.9J) and blackflies Simuliidae (fig. 4.9I), medial veins are located close to the Cu and altogether form a single stiffening rib of characteristic shape. This structure can be seen most distinctly in midges Chironomidae (fig. 4.9F), being a diagnostic feature of the family. Midges evidently possess one or several secondary veins. All the veins are brought together in pairs, with a thick pigmented membrane between them. Similar structures obviously have adaptive nature in larger insects, but the selective value of these structures in tiny forms is highly disputable. The opposite rather is true, because extra sclerotization makes the wings heavier and demands more energy. Thus, these structures are inadaptive in small insects.

Many advanced Muscida (as well as Vespida and Scarabaeida) are characterized by the reduction of costal vein fragments not observed in the Amphiesmenoptera. In flies, the rear part of the C is often absent, beginning from the place of its connection with the M or RS (fig. 4.3C, E, G–I; fig. 4.9D, F, I). Thus, the anal wing edge consists of membrane. The rear part of the C is also lost in many beetles and can even be absent on much of the leading edge (fig. 4.2B, C). The same is characteristic of hymenopterans (fig. 4.4). The reduction of the C is exhibited in the majority of, and also restricted to, insects with relatively high wingbeat frequency (see fig. 3.3). Accordingly, it can be assumed that costal reduction is caused by one of the two following selective factors:

1. Acceleration of flow over the wing surface leads to boundary-layer thinning, thereby enabling interaction between wing surface structures and the flow. The latter is most likely aerodynamically disadvantageous (see sec. 4.8.3); hence, natural selection favors the decrease of wing roughness in the parts where airflow is the most rapid, causing reduction of the rear part of the costal vein, so that microscopic projections of the wing surface do not stick out of the boundary layer or even its viscous sublayer. If this is true, the thickness of the boundary layer on flapping wings, which at present cannot be measured or calculated, can be estimated according to the size of the surface structure projections on an insect's wings.

2. The frequency increase leads to the growth of inertial loading, experi-

enced by wings at the points of trajectory reversal. Therefore, adaptive significance of the decrease of wing mass grows with stroke frequency and induces
the reduction of the costal vein as the bearing element of least importance.

From my viewpoint, the second cause is more likely, since the costal vein
can also be lost in wing parts that do not contact the airflow, for example, on
the rear edge of the forewings in hymenopterans (fig. 4.4). However, the relationship of the factors of reduction is still uncertain, and final conclusions
are not yet possible.

The primary furrows of dipterous flies are strongly or completely reduced
(fig. 4.3). The remigial furrow (fig. 4.3C) can be seen only in a minority of
species. A well-developed remigio-anal furrow can often be found near the
wing hinges (fig. 4.3A, E, G, H), but a completely developed p. remigio-analis
(fig. 4.3B) is quite rare. The convex ano-jugal furrow is present in all species,
but is very short and thus almost imperceptible, departing from the 3Ax.
The jugum is situated behind it, sometimes rudimentary (fig. 4.3A), sometimes modified and transformed into so-called upper and lower calypteres
(fig. 4.3D). Significance of the calypteres for flight is unclear. As already
noted, derivatives of the wing's jugal lobe also play no role in the flight of
diurnal butterflies (sec. 4.2.3) and hemipterans (Wootton 1996). Calypteres,
rather, are the morphologic consequence of the formation of the narrow wing
hinges.

The wings of dipterous flies are connected to the body by narrow (often
very narrow, rodlike) hinges, or stalks. This is the most probable reason for
the reduction of furrows and the decrease in wing deformation. The rear edge
of the stalk in many species carries a small lobe that can be found in *Aedes*
(fig. 4.9G), *Anisopus, Plecia,* and others. In the majority of the Muscida Brachycera this plate is increased, separated from the stalk by a broad fold line and
called the *alula* (fig. 4.3C–H). (Behavior of the alula during the stroke cycle
was described in section 3.9.) Presumably, the articulated alula had multiple
origins during the evolution of the Muscida. This can be ascertained by the
absence of the alula in part of Brachycera from different families: Stratiomyidae, Bombyliidae, Asilidae, Leptogastridae, Platypezidae; see, for example,
L. cervi L. (Hippoboscidae) (fig. 4.3I). Interestingly the structures participating in the control of airflow in four-winged insects are normally situated on
the rear part of the hindwings. Thus, the wings of Muscida simultaneously
bear both the fore- and hindwing features of four-winged insects.

Fenestrae in dipterous flies are mostly lost and can rarely be observed;

only some species have a thyridium (fig. 4.3*A, F*). Supposedly, this is caused by high stroke frequency, at which the zone of ventral flexion of the wing shifts proximally or even disappears, while inertia at stroke reversal is damped out by pronation and supination (Ennos 1988a). Consequently, many species have secondarily developed breaks in the costal vein. One, two, or three of them locate near the vein base (fig. 4.3*H*; fig. 4.9*D*) and are widely used in taxonomic diagnostics of family rank (Shtackelberg 1969). At the end of downstroke, the wing often flexes ventrally at these costal vein breaks. Nevertheless, it is difficult to determine wing flight deformation from allocation of the breaks. Flexion can be absent in forms with breaks, as in *Zophomyia temula* Scop. (Tachinidae), and vice versa, the wing can flex ventrally without any detectable breaks of the C, as in *Agathomyia viduella* Ztt. (Platypezidae) and *Antichaeta* sp. (Sciomyzidae) (Grodnitsky and Morozov 1994).

Ennos and Wootton (Ennos 1989b, 1989c; Wootton and Ennos 1989) describe an apomorphic adaptation of the Muscida to flight: the base of the medial vein is connected to the R and the CuA by a forklike joint of two short cross veins (as in fig. 4.3*C*). The joint is movable and forms an aerodynamically beneficial convex profile during the downstroke. However, data on dipteran wing deformation (Grodnitsky and Morozov 1994) do not corroborate this viewpoint. Furthermore, the base of the M does not camber even when a live or fixed wing is pushed with a pin.

While speaking of dipteran wings, the highly specific morphology of the wings of the mountain midges Deuterophlebiidae deserves mention. Larvae of these insects live in alpine creeks and the imagos swarm close to the water surfaces (Babcock 1985; Jedlička 1986; Turner et al. 1986). Their wings bear a well-developed complicated system of furrows (occasionally called *secondary veins;* see Jedlička and Halgoš [1981]). These furrows help plait the wings inside the puparium (Brodsky K. A. 1930); they do not operate in the imago. In addition, they apparently impart additional strength to the wings, so that venation becomes irrelevant and is strongly reduced (fig. 4.13*B*). The forewing folding system found in mountain midges is similar to that of earwigs (fig. 4.13*A*) and represents one of the most surprising examples of convergent similarity in insect wings.

4.5. Wing Morphology in Beetles

The version of vein homology on beetle wings suggested by Ponomarenko (1972) will be used in the following text. The more recent revision of venation

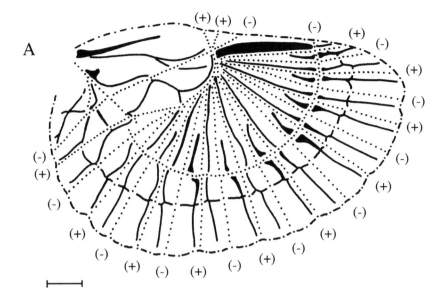

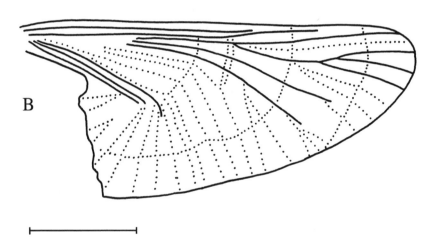

Figure 4.13. Wing morphology: (*A*) earwig *Anechura bipunctata* L. (Forficulidae); (*B*) mountain midge Deuterophlebiidae Gen. sp. nov. Designations as in figure 4.2. Scale 2 mm.

made by Kukalova-Peck and Lawrence (1993) is avoided, because it assumes that the initially weak medial vein has lost connection with the remigial furrow and traveled into the location occupied in all other insects by the anterior cubital vein. At the same time, the M became considerably stronger, whereas the CuA weakened and often even disappeared. Such a dramatic morphologic and functional change needs substantial additional support. At least a hypothetical transitional series of intermediate morphologic states from the ancestor of Scarabaeiformes to Scarabaeida is necessary. However, this and any other support for the revision have not been presented. Finally, the main query remains unanswered: Why should the direction of evolutionary transformation of venation common to all Scarabaeiformes (reduction of the proximal part of the M and the whole CuP, reinforcement of CuA) differ so profoundly in coleopterans?

Additionally, Kukalova-Peck and Lawrence (1993), without explanation, incorporate a convex anal furrow (anal fold) into the hypothetical construction of the primitive neopteran wing, believing that in the evolution of beetles it substitutes for the concave p. remigio-analis, whereas the convex p. ano-jugalis is reduced. These assumptions determine the corresponding changes in vein designations introduced in their paper. From my viewpoint, there is no evidence supporting the theory that the convex furrow separating the proximo-anal wing part that is folded up while plaiting is anything other than p. ano-jugalis. Correspondingly, the most powerful vein that passes along the rear border of the remigium can be nothing but the CuA. These designations are accepted in figure 4.2. Note, however, that beetle wings bear very diverse secondary sclerotization expressed as fields that are not shown in the figure. Homologization of the clavus veins is problematic, because they are very often weakened and morphologically identical to the false veins of other insects. Therefore, the structural scheme of this wing region has considerable individual variability. For example, 16 anomalies can be recognized in the venation of *Chlorophorus faldermanni* Fald.; 21 in *Xylotrechus namanganensis* Heyd. (Cerambycidae) (Chikatunov and Kriukova 1985); and 42 in *Entomoscelis adonidis* Pall. (Chrysomelidae), 18 of which are abundant (Chikatunov and Denisova 1988).

Ordinarily the size of the wings considerably exceeds the size of the elytra in the Scarabaeida. Hence, they have secondarily acquired furrows, along which the wings are folded up at rest (fig. 4.2). According to available data, this system of hindwing plaiting developed independently in the many groups of this order, resulting in extremely diverse folding systems. Furrows can therefore be homologized only after detailed comparative research. The available data (Ponomarenko 1972; Schneider 1978; Kukalova-Peck and Lawrence 1993; Zherikhin and Gratshev 1995) are insufficient for this purpose. Hence, the majority of furrows herein remain unnamed. Only p. ano-jugalis, always convex and departing from 3Ax, can consistently be recognized. To a lesser degree, this is true of concave p. remigio-analis. The remigial furrow and nodal line are difficult to identify, since within particular sections they change their sign and alter from concave to convex and vice versa.

Because they are corrugated structures, furrows increase wing strength. The system of furrows is constructed so that the wing possesses only two stable states, spread and folded up. Furrows impart such a stiffness to the wing that during the stroke cycle, the remigium (the zone within which most of the furrows are located) flexes neither in the longitudinal nor in the transverse direction (sec. 3.7). The ability of furrows to impart additional strength to the wing coincides well with the previously mentioned case of the Deuterophlebiidae wing morphology, in which the system of furrows partially substitutes for venation.

Any apparent correlation between the degree of anatomic expression of furrows and in-flight wing deformation is generally absent. For example, in *Monochamus urussovi* (Fisch., *M. sutor* L., *Corymbia rubra* L. (Cerambycidae), *Mordellistena tournieri* Emery. (Mordellidae), and *Cetonia aurata* L. (Scarabaeidae), the wings flex along the p. remigio-analis in the second half-stroke. In *Cassida* sp. (Chrysomelidae) and *Trichius fasciatus* L. (Scarabaeidae) they do not, although the furrow is well developed in all the mentioned beetles. On the other hand, in species having no anatomically expressed p. remigio-analis, clavus flexion is observed in *Cicindela restricta* (Carabidae) and in *Oedemera* sp. (Oedemeridae), and absent in *Coccinella trifasciata* L. (Coccinellidae) and *Lagria* sp. (Lagriidae) (Grodnitsky and Morozov 1994). This is also true of the configuration of the folded wing condition; the wing can be folded along lines that are imperceptible on spread wings. In general, a folded wing has considerably more lines of deformation than anatomically expressed fold lines, suggesting that wing plaiting originates first during evolution, followed by the development of specialized morphologic structures such as furrows.

4.6. Wing Morphology in Hymenopterans

Remigio-anal furrows are distinctly expressed on both the fore- and hindwings of Vespida. A long or short ano-jugal furrow (dependent on jugum size) is always present, although the furrow is practically imperceptible if the jugum is very small, as in the Muscida. The p. remigialis is not visible on either the fore- or the hindwings, even in the ancestral condition (Rasnitsyn 1969, 1980). Despite this absence, both wing pairs of most species of Vespida generally bend during upstroke at the sites where the p. remigialis and p. remigio-analis would pass (Grodnitsky and Morozov 1994). In some species, the remigial furrows become secondarily pronounced through thickening and pigmentation of the surrounding membrane. Examples include the hindwings of *Ceropales* sp. (Ceropalidae), *Acantholyda nemoralis* Thomson (Pamphilidae), *Pseudogonalos hahni* Spin. (Trigonalidae), and *Chrysis* sp. (Chrysididae), and both wing pairs of *T. femorata* F. (Tiphiidae) (fig. 4.4D). The forewings of *Tiphia* and *Chrysis* possess secondary furrows of unclear function.

Many species bend their wings along the convex nodal line. This line can be traced according to fenestrae on veins that usually are present and homologous to the thyridium and arculus of caddisflies and moths. Besides these two fenestrae, bullae on cross veins occur in the places where the veins intersect with invisible p. remigialis and visible p. remigio-analis. Bullae can also be located in places free of furrows. By contrast, the cross veins of *U. gigas* L. (Siricidae) are almost completely deprived of fenestrae (fig. 4.4B), although its flapping surface flexes along definite lines during flight (fig. 3.7A). Thus, in the Vespida, just like in other insects, frequent cases can be found of a discrepancy between morphology and flight deformation of the wings.

Plaiting of the flapping surface of higher hymenopterans is performed in a manner unusual for Scarabaeiformes. A single secondary convex furrow is situated on the forewing. Unlike in any other insect group, this furrow has no connection with 3Ax. This peculiarity probably permits the furrow to change the sign of its profile from convex to concave during upstroke (Brackenbury J. H. 1994). Polarity change is extremely rare in insect wing furrows and is generally prohibited by mechanical properties of the wing hinge.

In addition, the concave fold line corresponds to fore- and hindwing coupling or, in Pompilidae, to p. remigio-analis. Danforth and Michener (1988) have described similar plaiting in detail for several families of wasps, as well as for bees *Eulonchoptera* (Colletidae). They have also derived a direct correlation between the presence of the convex furrow and the fenestra on the last cross vein situated between the M and CuA. The latter is incorrect, as some hymenopterans, for example, *Rhogogaster* sp. (Tenthredinidae) (fig. 4.4A), possess a bulla on the indicated vein although the wings do not fold up at rest.

Tobias (1992, 1993) considers natural selection to be directly responsible for transformations of venation observed in the Ichneumonoidea, Sphecoidea, and Apoidea. Within these groups, species that live in open, arid biotopes have wing veins that are shifted proximally. Tobias (1992, 1993) suggests that strong winds eliminate weakly flying specimens and thus improve the flight performance of inhabitants of open spaces. This interpretation is disputable, because in dipterous flies just the opposite tendency is observed: perfection of their flight is accompanied not by a proximal, but by a distal shift of wing venation (Ussatchov 1970; Stary 1990). On the other hand, Rasnitsyn (1969) suggested that veins shift toward the base and leading edge of wings (this process is called *costalization*) under a decrease in absolute size of specimens. This hypothesis has not been falsified and also deserves mentioning.

4.7. Morphology and Functions of Furrows

Insect furrows are more diverse than veins. Venation inherited from ancestors can change in evolution within some limits. In most cases the changes are related to oligomerization, that is, reduction of some veins and reinforcement of others. Polymerization of venation reported for archaic groups (Rasnitsyn 1980) has rarely been observed in advanced taxa. Ordinarily only false veins are formed secondarily; a lost vein never returns. If natural selection demands strengthening of a particular wing zone, it is achieved by means of corrugation, thickening, and sclerotization of the membrane, unaccompanied by formation of new true tubular veins. Veins can develop only from tubular vestiges available in ancestors, as they do in the case of jugal venation in many taxa (sec. 4.2.3; Martynov 1924b; Snodgrass 1935; Shwanvich 1949; Sharov 1968; Rasnitsyn 1969).

The following types of furrows can be recognized:

1. Anatomically expressed lines of flight deformation. This type of furrow can be seen on the spread wing, because it differs from the surrounding membrane by its relief and/or color owing to desclerotization or lesser thickness. Anatomically expressed furrows include the primary concave furrows: p. remigio-analis in hymenopterans; p. remigialis and p. remigio-analis in the majority of caddisflies, moths and butterflies.

2. Invisible (not expressed anatomically) lines of in-flight deformation: nodal line in all insects except beetles; p. remigialis in hymenopterans; the line of trapezoid deformation in caddisflies and moths. The arrangement of furrows of this type can be traced by the arrangement of the microtrichia, which often have another size or form on the furrow, differing from the microtrichia on neighboring plots of the wing. Often (but far from necessarily) furrows pass through breaks in the veins, that is, bullae (or fenestrae, which is the same).

3. Anatomically expressed lines of bending that work at plaiting. In these furrows (it is unknown how consistently), inner tension exists that folds up the wing at rest. The ano-jugal fold line in all insects is a typical example of this furrow type, as are many of the secondary furrows present in beetles, caddisflies, and moths.

4. Invisible lines of bending at plaiting: secondary convex furrows on forewings of hymenopterans. These furrows probably do not have inner tension because they can change their polarity (see sec. 4.6).

5. Peripheral furrows. These structures do not participate in wing deformation but instead represent a type somewhat different from the previous four. They should probably be designated by a term other than "furrow."

In all likelihood, the first four types of furrows are interconvertible. Thus, in the majority of the Amphiesmenoptera, p. remigialis belongs to type (1), in pyraloid lepidopterans to (3), and in hymenopterans to (2). Again, p. remigio-analis on hindwings is generally type (1), whereas in the caddisfly family Phryganeidae, in some Calamoceratidae, in moths Arctiidae, Pyralidae, Pyraustidae, Crambidae, and in the honeybee (fig. 4.4C) it is type (3).

Therefore, the term "furrow" corresponds to at least five morphologically and functionally different structures. This is the first and, hence, a quite hesitant attempt to build a furrow typology. According to the general considerations (sec. 4.2.1, see also under "Multivariate Correspondence between Structures and Functions" in the appendix), classification must be based either on morphologic or functional characters. In the case of furrows, these can include features of surface microstructures or inner morphology of the folded membrane, its thickness, and presence of elastic proteins. Unfortunately, no comparative survey on furrow morphology has ever been undertaken, let alone "conducted."

Furrows, like other membranous structures, originate in the course of evolution quite easily. This can be seen from the basic difference of the considered systems of wing plaiting found in different groups of endopterygote insects. Moreover, in each order, not one but several different methods of plaiting can be found. Finally, an impression can be derived that during insect wing evolution, a change of adaptive strategies took place, because polymerization of venation has been substituted by the development of new furrows.

It is difficult to unequivocally indicate why furrows developed more easily than veins. Furrows are possibly more functionally adjustable than veins as factors of changeable corrugation. A vein can serve solely as a stiffening element of construction. A furrow (e.g., longitudinal) enables deformation of the wing profile, simultaneously preventing the wing from bending in the baso-apical direction. Furrows also weigh considerably less than veins. In addition, a system of furrows of definite configuration, as occurs in beetles, causes greater wing stiffness, which cannot be provided by veins, unless they are thick enough.

Thus, each furrow can play several roles, causing in-flight deformation of wings, their plaiting at rest, and an increased wing stiffness or resistance to flexible deformations. In various insects, each role is fulfilled by a different combination of furrows and fenestrae and can sometimes be satisfied without both structures. Consequently, furrows and their functions appear to be connected by multivariate correspondence (fig. A2B).

4.8. Lepidopteran Wing Scale Covering

The presence of a solid covering of scales constitutes the most characteristic feature of the Papilionida order. To a great extent, the covering has conditioned and channeled the evolutionary transformations of moth and butterfly wings, preventing an increase in stroke frequency. Types of covering are related to the development of complex mechanisms of thermoregulation that admit two alternative adaptive strategies (sec. 3.8). Meanwhile, the covering has obviously been studied insufficiently. Only ultramicroscopic morphology (Onslow 1921; Schmidt and Paulus 1970; Kristensen 1978) and shape variability (Müller 1972) of scales, including androconia (Sellier 1971, 1972, 1973a, 1973b; Niculescu 1978), have been investigated. Organization of the covering as a whole, although clearly significant for the evolution of lepidopterans, has never been discussed. Nevertheless, some features of the covering can be used in taxonomy (Grodnitsky and Kozlov 1989b, 1990a). The morphology and possible functions of the scale covering in the Papilionida will now be considered.

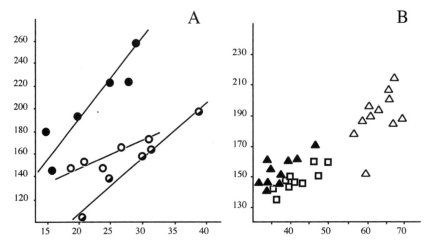

Figure 4.14. Correlation of scale length (vertical, m · 10⁻⁶) and wing radius (horizontal, mm). (*A*) Different species of Satyridae (●), Nymphalidae (○), and Sphingidae (◑); (*B*) different specimens of swallowtail butterflies *Iphiclides podalirius* L. (▲), *Papilio machaon* L. (□), and *Papilio maacki* Men. (△).

4.8.1. Size of Scales

In the majority of moths and butterflies, scales have the same size in both proximal and distal wing zones (Grodnitsky and Kozlov 1989a, 1989b). Scale length normally displays a direct relationship with wingspan in both inter-specific (fig. 4.14*A*) and intraspecific comparisons (fig. 4.14*B*): mean length of the scales grows with an increase in wing length. This correlation is absent only in swallowtail butterflies of the genera *Parnassius* and *Zerynthia* (Grodnitsky and Kozlov 1989b).

4.8.2. Covering Structure and Types

Organization of the scale covering is characterized by the distribution of morphologically different scales within the wing surface, by the pattern of their arrangement and regularities of orientation as related to location on the wing.

4.8.2.1. Orientation of Scales

Orientation of the scale tips is uniform in primitive moths, since all the scales are situated along the baso-apical axis of the wing and only decline slightly toward the nearest edge. In ghost moths, the wings possess a well-expressed tornal angle, and the scales are oriented in a fanlike manner, crossing veins

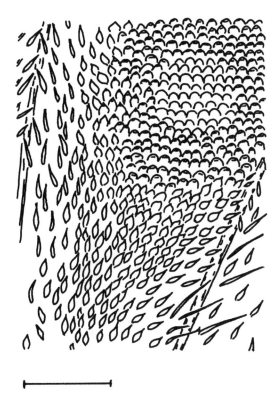

Figure 4.15. Arrangement of scales within eyelike pattern on lower surface of *Parnassius* sp. hindwing. Scale 1 mm.

and overlapping furrows at different angles. In tiny lepidopterans, the orientation of scales is analogous to that of Micropterigdae and Eriocraniidae. The direction of scales in large forms coincides, as a rule, with the direction of the nearest veins. Only the scales that are adjacent to a vein are situated under an acute angle to it, covering the vein with their vertices. The scales can overlap subcostal and radial veins under small angles within the costal field of the lower wing surface, orienting toward the wing tip (Grodnitsky 1988).

Elements of a wing's color pattern can control the initial orientation of the scales. For example, in the field of eyelike patterns displayed on the wings of swallowtail butterflies of the genera *Archon* and *Parnassius*, elongated scales surround the core of the pattern, departing prior to it and then meeting again (fig. 4.15).

4.8.2.2. Rows of Scales

Scales in primitive lepidopterans are not ordered, a state that can be considered primitive for the order. In the majority of butterflies, the scale sockets as well as the scales themselves are arranged in tranverse rows. The rows of neighboring cells of the wing come to the vein that separate them under obtuse angle, so that the rows on the entire wing are oriented along secants of concentric circles with the wing hinge as the center (Yoshida et al. 1983). This orientation pattern is violated only in bands of one to three scales, adjoining veins and wing margins, where the scales are longer and their tips are to the side of the vein (or wing margin). The carpenter moths Cossidae, whose scales and scale sockets are not put in order (Grodnitsky and Kozlov 1989a), provide a unique exception.

Scales do not form transverse rows in small narrow-winged moths. Such an arrangement can be primitive or secondary. Scale sockets in lappet moths are arranged in order, but the scale tips appear to be disordered owing to a size polymorphism of the scales (Grodnitsky and Kozlov 1990a).

Within a family the covering can be ordered to various degrees. For example, the rows are expressed distinctly in swallowtail butterflies of the subfamily Papilioninae. In the subfamily Parnassiinae, the covering is more diverse: in *Bhutanitis thaidina* Bland, the rows are as well expressed as in Papilioninae; in *Serecinus telamon* Don., *Allancastria cerisyi* God., and *Luehdorfia puzioli* Ersch., they are less distinct; and in the tribe Parnassiini, the scales are not arranged in rows, and even the scale sockets are scattered chaotically on the wing membrane (Grodnitsky and Kozlov 1989b).

However, note that among swallowtail butterflies, the transverse rows of scales are well expressed in species with long tails on the hindwings (*Bhutanitis, Serecinus, Papilio, Parides, Iphiclides*); exceptions include some species of *Papilio* and all species of *Ornithoptera*, which have rows but not tails. The rows are much less distinct in *Allancastria* and *Luehdorfia* with small tails. Species of the tribe Parnassiini possess neither tails nor rows. The covering of the tails themselves does not differ from the covering of the rest of wings (Grodnitsky and Kozlov 1989b).

4.8.2.3. Covering Layers

The terms *one-*, *two-*, and *multilayer covering* will be used to describe the inner structure of the wing covering (Kristensen 1970, 1974, 1978; Nielsen and Davis 1981, 1985; Kozlov 1987). The curves of scale length distribution are suitable for a more exact description. One- or single-layer covering is characterized by uniform scales, a distribution described by a curve with expressed positive excess (fig. 4.16A). A two-layer covering is composed of scales of two morphologic types; the length distribution is bimodal (fig. 4.16B), because

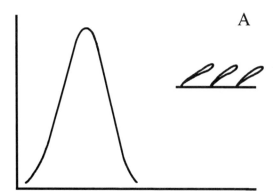

A

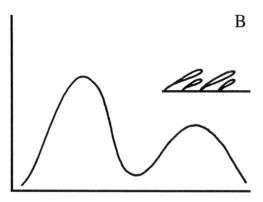

B

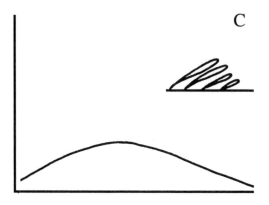

C

Figure 4.16. Modifications of scale covering structure: (*A*) one-layer; (*B*) two-layer; (*C*) multilayer. Horizontal, length of scales $(m \cdot 10^{-6})$; vertical, percentage of scales of given length in common sample.

intermediate forms of scales are scarce. The covering is termed *multilayer* if a continuous series of variable scales is present and the distribution curve has significant negative excess (fig. 4.16C). Note, however, that these are strictly terms of convenience, since the covering has no layers in the direct meaning of this word.

The covering of primitive lepidopterans is characterized by a state that is the most primitive within the order (Kristensen 1970). The wings of *Micropterix calthella* L. (Micropterigidae) are covered by uniform scales, among which larger and hairlike scales can be found. The forewing of *Eriocrania semipurpurella* Stph. (Eriocraniidae) bears the same covering as mandibulate moths, but its hindwings are covered with scattered, narrow elongated scales.

The shape of the distribution of scale lengths in the majority of lepidopteran taxa is similar to that in primitive moths. The covering structure does not vary, or only varies slightly, for a particular wing region. Furthermore, primitive groups already display specialization of the covering. For example, scale distribution in the ghost moths Hepialidae is described by an excessively peaked curve, and distribution is bimodal in the fairy moths Adelidae. This demonstrates that different modifications of the wing covering have repeatedly originated in the course of evolution of the order. The suggested scheme (fig. 4.17) reflects the most likely routes of the covering transformation.

The decrease in scale size polymorphism observed in the Hepialidae, Pieridae, and Papilionidae corresponds to the formation of a single-layer covering. By contrast, the increase in scale variability leads to development of a multilayer covering, with the curve of distribution becoming stretched along the horizontal axis. Such a covering structure characterizes most species belonging to the superfamilies Pyraloidea, Bombycoidea, Hesperioidea, and, partly, Papilionoidea (Lycaenidae). The covering of the Geometroidea reflects an intermediate stage of the development of multilayer covering.

Intermediate scales are reduced in the Nymphalidae, Satyridae, and the majority of the Lasiocampidae, and the covering is formed by the scales of two size categories. Having equal functional capacities, such covering has less mass and, consequently, reduced wing inertia.

The process of two-layer covering origination can be traced using the families Lasiocampidae and Thaumetopoeidae (tab. 4.1; generic names according to Fletcher and Nye [1982]) as an example. The multilayer covering typical of the bombycoid complex reflects the initial state for both families. It has been found only in *Eriogaster lanestris* L., while close organization was

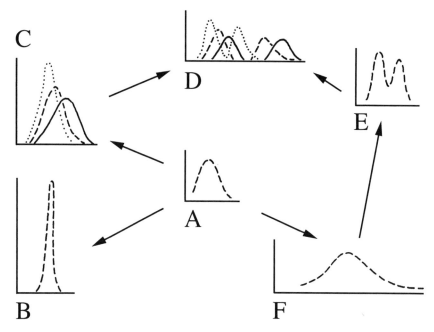

Figure 4.17. Interrelationships among different types of scale covering. Scales taken from wing base (solid line), midspan (dashed line), and wing tip (dotted line). Coordinates as in figure 4.16.

observed in *Thaumetopoea herculeana* Rbr. A hypothetical process of the evolution of the lappet moth wing covering can be divided into the following stages.

The first stage consists in the partial reduction of scales of intermediate length: the length distribution curve acquires a second maximum. Distribution of the number of teeth remains unchanged and is described by a one-vertex curve (tab. 4.1; *Amuriella dieckmanni* Graeser, *Paralebeda plagifera* Walk., etc.).

Next, partial reduction of the scales of intermediate length is accompanied by morphologic divergence between large and small scales: the curve of teeth number distribution also becomes two-vertex (tab. 4.1; *Gastropacha quercifolia* L., *Arguda vinata* Moore, etc.). In these species the wing covering consists of large scales with one to three teeth and small scales with four to eight teeth. Scales with few teeth are almost fully hidden under scales with more numerous teeth (Grodnitsky and Kozlov 1990a), that is, the covering is divided into overlapping and underlying parts. The long teeth of the underlying scales facilitate intercalation of air between the layers.

TABLE 4.1
Modifications of wing scale covering in lappet moths Lasiocampidae and Thaumetopoeidae

Scale-length distribution[b]	Teeth Size[a]		
	Small	Pronounced	Extreme
l / t	Hypothetical modification primary for the family		*Eriogaster lanestris*
l / t	*Thaumetopoea herculeana*	*Amuriella dieckmanni*	*Paralebeda plagifera* *Suana concolor* *Macrothylacia rubi* *Amuriella fulgens* *Metanastria nanda*
l / t	*Gastropacha quercifolia*	*Eustaudingeria vandalica* *Arguda vinata* *Trichiura crataegi* *Metanastria hyrtaca* *Philudoria potatoria* *Philudoria albomaculata* *Trabala vishnou* *Gastropacha populifolia* *Odonestis pruni*	
l / t	*Malacosoma franconica* *Malacosoma neustria* *Malacosoma castrensis* *Dipluriella loti*	*Chilena sordida*	*Eriogaster catax*
h / l / h / t / l	*Thaumetopoea solitaria* *Thaumetopoea pityocampa*	*Phyllodesma ilicifolia* *Phyllodesma glasunowi* *Phyllodesma suberifolia* *Phyllodesma alice*	*Nadiasa undata*
l / t			*Poecilocampa populi* *Syrastrenopsis moltrechti* *Bharetta cinnamomea* *Cosmotriche lunigera*
h / l / h / t / l	*Thaumetopoea iordana*		

[a] Depth of incisions on the scales.

[b] Degree of differentiation of the covering into two layers, according to the shape of distribution of scale lengths (l) and the amount of teeth on the scales (t); h, hairs substituting for the upper scale layer.

The differentiation of the two layers is then further enhanced: the maxima on the curves of the distribution of teeth number and scale length become more distinct; and intermediate scales become considerably rarer (tab. 4.1; *Malacosoma franconica* Esp., *Chilena sordida* Ersch., etc.). In *Malacosoma neustria* L., *C. sordida* Ersch., and *Eriogaster catax* L., the intermediate scales contribute up to 9–12% of the common sample, whereas in *G. quercifolia* L., *Trichiura crataegi* L., *Metanastria hyrtaca* Cramer, and in other species characterized by the previous stage of covering differentiation, the intermediate scales contribute up to 15–25%.

Complete morphologic differentiation of the covering is achieved with the replacement of the large scales of the upper layer with hairs: underlying scales are small, and the overlapping layer is presented by hairs (tab. 4.1; *Thaumetopoea solitaria*, *Phyllodesma ilicifolia* L., etc.).

During the next stage of the evolution of covering, one of the layers is reduced. Thus, in *Thaumetopoea iordana* Stgr. scales on wings are scarce, constituting no more than 25% of the covering, and the covering itself is basically formed by hairs, as in the majority of caddisflies. By contrast, in some species hairs are reduced (tab. 4.1; *Poecilocampa populi* L., etc.); the covering consists of scales with very long teeth and, therefore, looks practically identical to the caddisfly-type covering. Distributions of teeth number and scale lengths in these groups are described by one-vertex curves. Both cases of reduction produce the same external result: the wings of the moths become half-transparent. Note, however, that blank cells in table 4.1 indicate that there can exist species of lappet moths with a covering structure corresponding to empty classes.

4.8.2.4.   Differentiation of the Covering as a Function of Wing Zone

In the superfamilies Noctuoidea, Notodontoidea, Sphingoidea, and Lasiocampoidea, the development of a multilayer covering is accompanied by its differentiation within the wing surface. The flat scale-size distribution curve indicates a multilayer structure of the covering near the wing base. Farther from the wing articulation, the covering is formed by more uniform scales and gradually acquires a single-layer structure, and the curve displays a narrow peak (fig. 4.17C). Additionally, unlike in other groups of lepidopterans, the mean scale length decreases toward the edge of the wing surface. Simultaneously the angle of inclination of the scales to the wing membrane noticeably changes. Therefore, the total wing thickness (including scale covering) of the moths Noctuoidea, Notodontoidea, Sphingoidea, and Lasiocampoidea

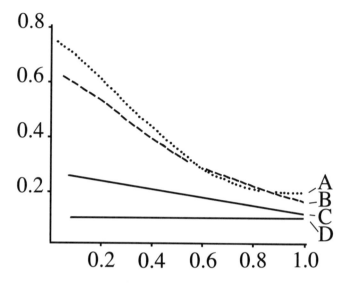

Figure 4.18. Correlation between total thickness of wing scale covering (vertical, mm) and nondimensional wing span (horizontal): (*A*) hawkmoth *Laothoe populi* L. (Sphingidae); (*B*) noctuid moth *Catocala nupta* L. (Noctuidae); (*C*) measuring worm moth *Arichanna melanaria* L. (Geometridae); (*D*) butterfly *Boloria dia* L. (Nymphalidae).

decreases considerably in the base-to-tip direction. By contrast, the wings of the Geometroidea, Pyraliodea, Hesperioidea, and Papilionoidea are almost uniform in thickness (fig. 4.18).

In some lappet moths, the covering of some wing parts differs by the degree of differentiation into overlapping and underlying scales. Thus, in *C. sordida* Ersch., scale length and number of teeth distributions are described by a two-vertex curve in any zone of the wing (fig. 4.17D). The covering of *G. quercifolia* L. is also subdivided into two layers in any forewing region, although close to the external margin both layers are presented by two- to four-teeth scales, whereas at the wing base and discal cell the overlapping layer consists of one- to two-teeth scales, and the underlying scales are three to six teeth. The most significant differences in the construction of covering in different wing areas are observed in *M. hyrtaca* Cramer. Within the proximal third of the wing the covering is two-layer, and the underlying and overlapping scales differ in size and shape. Closer to the distal third of the wing the covering becomes increasingly uniform: near the outer margin it consists of scales whose size and shape are described by one-vertex curves, as in noctuid moths, prominents, and hawkmoths.

The scale covering of the fore- and hindwing probably did not differ initially. However, even in mandibulate moths the scales are narrower and scarcer on the hindwings than on the forewings. The hindwing covering is also somewhat different in some eriocraniid moths (Grodnitsky and Kozlov 1985). In many brush-footed butterflies, the covering on the forewings is two-layer, whereas their hindwings bear a single-layer covering in the main part and a two-layer one (with inclusion of opalescent scales) on the ano-jugal lobe that adjoins the abdomen. According to available data, most lepidopterans have a scarcer covering of scales on the hindwings than on the forewings.

The coverings of the upper and lower wings' sides also differ. In primitive moths (Micropterigidae, Eriocraniidae, Psychidae), the lower side of the forewings is covered by considerably narrower scales than the upper. The lower wing surface of *Parnassius* is covered by scattered hairlike scales, except in the region that has a colored pattern. The distribution of scale lengths can also display differing patterns. The upper wing surface of hawkmoths has a covering in the central region similar to that near the wing tip; on the lower surface, the covering of the central zone closely resembles that of the proximal part of the wing.

In functionally two-winged lepidopterans, parts of the forewing's lower side and hindwing's upper side overlap in flight. The covering of these parts generally differs from neighboring zones by scale size, shape, and orientation. In the overlapping zone, scales are never arranged in transverse rows. Scale covering of this wing part is denser; scales on the forewing lower surface often acquire a lanceolate shape and orient toward the PCu rather than toward the wing edge. The scale covering of the hindwing leading edge also becomes denser, although the scale orientation is unchanged. These peculiarities of scale covering apparently help couple the wings by acting as a coupling mechanism. Note, however, that similar transformations of covering (change of orientation, shortening and relative thickening of microstructures) in the regions of overlapping of the fore- and hindwings also occur in caddisflies, for example, *P. bipunctata* Retz., *Oligotricha striata* L. (Phryganeidae), *G. pellucidus* Retz., *A. laevis* Zett., *C. villosa* F., *Parachiona picicornis* Pict. (Limnephilidae), *C. nigronervosa* Retz. (Leptoceridae), and many others.

The covering often undergoes significant differentiation in respect to color patterns. Thus, structural coloration of the wings of lycaenid butterflies changes the primary ground plan of the covering construction. The blue coloration of the upper side of male wings of *Polyommatus coridon* Poda, *Polyommatus amandus* Schn., *Polyommatus daphnis* Den. et Schiff., and *Celastrina argiolus* L. is formed exclusively by small rounded scales, whereas large scales are nearly absent in this area. For example, in *P. coridon* Poda, the scales of the blue spots are rounded, whereas the brown pigment pattern adjoining the wing margin and veins is formed by small, intermediate, and large scales, like the covering of the lower wing surface (Grodnitsky 1988).

The covering within pigmented regions of the forewings becomes multilayered in some species of Papilioninae (*Papilio ormenus* Guer., *P. ulysses* L., *P. penelope* Wall., *P. ybecatheus* Wall.); these parts of the wings do not display the arrangement of scales in rows generally typical of the subfamily. Strong changes of the covering structure are seen in the eyelike pattern in the Parnassiinae (fig. 4.15). Light spots on the wings completely lack scales and hairs in some of the Syntomidae (*Lethe eumolphus* Ev., *Melanthus* sp., etc.), and the spots themselves may be colored white or yellow by pigment contained in the wing membrane.

4.8.3. Functions of the Covering

The experimental removal of scales from the body and wings of lepidopterans has demonstrated that the scale covering possesses pronounced heat-insulating properties, equivalent to mammalian hair, avian feathers, or the best man-made heat-insulating materials (Church 1960; Heinrich 1970, 1993; Heinrich and Bartholomew 1971; Ivlev 1993a, 1993b). The heat-insulating capacity of this covering is best manifested on the pterothorax and at the wing hinges. The presence of the scale covering results in a 1.5–2-fold difference between the temperature of the thorax and that of the surrounding environment (Church 1960). Some lepidopterans, like many other insects, are able to maintain a high (up to 37–45°) body temperature, necessary for the normal energy consumption of the pterothoracic muscles during flight (Heinrich and Bartholomew 1972; Ushatinskaya 1987).

The specialization of lepidopterans at either ecto- or endothermic temperature regulation influences the structure of the wing covering, as well as the wing shape (sec. 3.8). Taxa with generalized wings and comparatively weak flight possess a uniform scale covering across the entire wing surface, and the distribution of scale lengths is described by the curve of intermediate type. In typically endothermic lepidopterans (Sphingoidea, Noctuoidea, et al.), the construction of the covering changes from the base to the tip of the wing. A thick covering of the basal part prevents strong heat loss through broad veins and hemolymph vessels. A thin one-layer covering at the wing tip is associated with much less (as compared to the proximal part) heat exchange in this area (Heinrich 1993) and with the necessity to reduce the covering mass, as the momentum of inertia grows with increasing distance from the wing hinge.

Three factors may contribute to the near absence of a scale covering on the wings of some endothermic forms with a dirunal mode of life (Aegeriidae, some Sphingidae). First, these particular lepidopterans are characterized by the highest stroke frequency in the order (sec. 3.8). This regime of thoracic muscle contractions in the warm daytime may potentially cause overheating and consequent flight termination in these moths (Heinrich 1970, 1993). In such a situation, reduction of scale covering permits muscle cooling during active flight. Second, a higher wingbeat frequency results in an increase in energy wasted in overcoming wing inertia at the upper and lower points of the stroke trajectory, when the insect needs to decelerate its flapping surface and then accelerate it again in the reverse direction. The loss of scales reduces the wing mass and, therefore, the total energy cost of flapping flight as well. And finally, covering reduction may aid in mimetic coloration (e.g., imitation of wasp appearance).

Construction of the covering of ectothermic lepidopterans, as a rule, is independent of the wing region. The one-layer covering becomes more complex in connection with color pattern elements only in some species. The covering characteristic of skippers can apparently be considered primitive for butterflies. In the course of evolution, it became transformed in two major directions. In the Nymphalidae and Satyrdae, the scale covering differentiates into two layers. This two-layer structure probably originated as a further specialization of the covering in its role as a heat insulator. The melanistic covering of satyrs and brush-footed butterflies favors even better heating of the butterflies' bodies (Watt 1968; Kayser 1985).

The single-layer wing covering of the Pieridae presumably does not possess good heat-insulatory capacities. The next stage of evolution of this type of covering is observed in *Parnassius* Latr.: the scales are so scattered that the covering is obviously unable to effectively protect the body and wings of the butterflies from cooling. Apparently, pierid and swallowtail butterflies possess biochemical, physiologic, or behavioral peculiarities (see the review by Heinrich [1993]) that compensate for the loss of the warm insulating function of the scale covering.

Numerous hypotheses on the aerodynamic significance of scale covering have been offered. Scales undoubtedly interact with the air and somehow affect the condition of the boundary layer. However, the specific mechanisms and the degree of this effect are still unknown. For example, according to Nachtigall (1965, 1967; see also Bocharova-Messner 1979b), scales turbulize the boundary layer and in this way improve lift-to-drag ratio of the wings. Brodsky (Brodsky and Vorobjov 1990), actively supports this viewpoint departing from the experimental data of Vorobjov (1991), who has shown that the increase of roughness on the lower side of a mechanical wing leads to partial deceleration of flow under the wing and hence to the growth of aerodynamic force.

Nachtigall's hypothesis demands that microscopic projections created by the covering fall outside the viscous zone of the boundary layer. Now let us compare lepidopterans with other flying and swimming animals and aircraft. The extreme size of projections of roughness on aircraft wings, at which elements of microrelief still remain within the viscous sublayer, constitutes 0.01–0.1 mm (Schlichting 1974). Although the roughness of lepidopteran wings is close in value, the Reynolds numbers that characterize the functioning of lepidopteran wings (about 10^3), are several orders less than the Reynolds numbers of aircraft profiles.

Next, the hydrodynamically active scales of fast-swimming sharks are 0.2–0.5 mm in size (Bechert et al. 1985), that is, several times larger than the projections of scales, even when they are inclined to the wing membrane at an acute angle. In addition, the Reynolds numbers of swimming sharks are also much higher, and the boundary layer of the sharks is much thinner than that of lepidopterans, owing to the difference in size between these animals and also the relative viscosity of air and water.

Additionally, it is well documented that hydrodynamically effective surface structures in animals always decrease with the growth of body size (Kozlov 1983); that is, the dependence of scale size on wing span, in contrast to the presented observations (fig. 4.14), must be inverse. Reasoning from the presented considerations, it can be suggested that the rough calculation of boundary-layer thickness by the formula for stationary flow (see sec. 2.7) gives a correct approximation: the scales do not come out of the viscous zone and do not affect airflow. The scales (like any microstructures) enlarge the surface of contact between the air and wing and, therefore, probably prevent flow separation. This does not, however, support the hypothetical role of the covering as a turbulizer, although the influence of larger structures (hairs and veins) on the flow cannot be excluded.

Generally speaking, the survey of lepidopteran wing scale covering morphology reveals its lability in response to different functional demands. This was illustrated by the adaptation of the covering to heat insulation, to reducing wing inertia at the increase in stroke frequency, and to the necessity of support of fore- and hindwings coupling. Most evident are the previously described changes in the covering structure within different elements of color pattern. Characteristically, almost all pattern-conditioned anomalies of the covering are observed within the distal half of the wing, the region that does not participate in thermoregulation. Given these facts, adaptations to the improvement of wing flight properties are at least not evident. For example, one could anticipate that scale orientation would coincide with the direction of airflow near the wing surface. However, comparison of morphologic data with information on flow around flapping wings (sec. 2.4–2.6) does not reveal such a correlation. Therefore, the assertion that scales do not interact with free airflow seems to be the least contradictory to currently known facts.

Thus, construction of the scale covering, to a considerable degree, is defined by the requirements of heat insulation and color pattern. Additional functions of scales can include aerodynamic (but not in the sense of Nachtigall's hypothesis), autotomic (scales permit moths to escape from spider webs and avoid entrapment in drops of plant resin), and behavioral, for example, signaling in courtship (Ono 1977, 1980). Meanwhile, no functional explanation exists for the origin of scales themselves. Scales are found on the wings and bodies of numerous insects besides lepidopterans, including beetles (Curculionidae), dipterous flies (Culicidae), hymenopterans, and caddisflies, and they are even seen in the Enthognatha (Collembola) and Thysanura (Lepismatidae, Machilidae) (Shwanvich 1949), always as secondary structures. Scales do not form a continuous covering in all the previously mentioned taxa, unlike in lepidopterans, so their respective functions are unclear.

4.9. Conclusions

The presented data and analyses demonstrate that modern knowledge on flapping flight aeromechanics (chap. 2) can quite effectively explain the initial radiation of existing types of flight apparatus and wingbeat kinematics, underlying the historical development of aerial locomotion of insects (chap. 3). Furthermore, thorough comparative analysis of wing morphology in endopterygotes (chap. 4) indicates a series of structural similarities of insect wings that cannot be explained by any similar functions fulfilled by these structures.

Of course, this situation can be ascribed to the lack of our knowledge of the functions of insect wings, since we always can hope to learn tomorrow what we do not know today. However, this method of interpretation does not seem sufficiently reliable to me owing to the following circumstance. We cannot anticipate that, from a general biologic perspective, the knowledge of adaptive functions (however substantial it may be) would provide us with an exhausting method of explanation of morphologic features found in living organisms. Every time an organism must change, it possesses just a limited set of alternative developmental trajectories, despite current demands of changing environment. These types of limitations are called *developmental constraints*. They are ubiquitous; the influence of developmental constraints can be neglected only in studies of organisms that lack ontogeny per se. They can be, for example, bacteria, which must have a generally larger percent of adaptive traits. However, there are no reasons to predict the same outcome for insects and their wings.

Developmental constraints originate from the self-organizing collective behavior of embryonic cells and from the interaction of different vestiges in the course of individual development. The latter can be additionally illustrated by the perspective of development of the vertebrate limb presented by Oster et al. (1988). The limb skeleton develops through the condensation of mesenchymal cells that form cartilaginous vestiges of prospective digits. Only three types of cell condensation are possible: focal condensation, branching bifurcation, and segmental bifurcation. Imagine that the conditions in which the limb operates were to favor trifurcation (i.e., branching of one element into three or more parts), leading to a more adaptively advantageous limb skeleton. Such morphology could not be realized, since trifurcation is only possible under a very narrow set of developmental parameters and therefore is extremely unlikely (Oster et al. 1988). However, the organism would not inevitably reach extinction. Adaptation to external environment bears the character of adequacy (Gans 1993): organisms should be just "sufficient" in relation to the conditions in which they live, whereas the concept of optimality—which has been repeatedly announced to be a reliable basis for calculating morphologies from functions—has quite a limited number of successful applications (see under "Optimality and Adequacy" in the appendix).

Summarizing the given example, we must admit that we do not have data indicating embryonic reasons for the development of particular homoplastic features of insect wings discussed in this chapter. Furthermore, all available data on the causation of morphologic peculiarities of living beings indicate that to reliably derive morphologies directly from functions, based on the concept of functional optimality of structures, is impossible.

These considerations can be additionally supported. Speaking of animal morphology in general, even the most widespread functional explanations of particular morphologic features are found to be untrue after considering comparative data. For example, we know that the melanic form of the birch inchworm moth received a strong selective advantage in comparison to gray specimens after the lichens growing on tree bark were killed by air pollution in nineteenth-century England. The advantage of the dark coloration was demonstrated in experiments in which birds easily found and ate gray moths on the dark bark deprived of lichens (Kettlewell 1955). The selective basis for industrial melanism was considered to be proved and was subsequently incorporated into numerous textbooks on evolution. However, industrial melanism is also found in unpalatable insects (e.g., ladybirds) not connected in their life with tree bark and lichens; moreover, melanism can be observed even in cats (Krassilov 1986). Evidently, melanism results from a far more general reason than that reported for birch inchworm moths.

Another example, also ubiquitous in textbooks, is given by the Irish elk *Megaloceros*. The immensity of its antlers is usually attributed to a functional advantage under sexual selection, specifically competition between males for females. The winning males presumably had larger antlers, which eventually became so large that they were strongly malfunctional in situations besides courtship behavior, and finally caused extinction of the elk. In fact, *Megaloceros* antler size closely fits the positive allometry of skeleton structures of all the family Cervidae (Gould 1974, 1977), which means that the antlers were this large simply because the elk itself was quite large. The size increase in these animals could have resulted from an entirely different cause—segregation in size owing to competition with related forms seems to be one of the most common methods of adaptive radiation within ecologically uniform groups, for instance ungulates (Kokshaysky 1988a).

A third widespread example concerns the eyelike patterns on insect wings.

Their functional role in butterflies is usually interpreted as for deflecting bird attack, a viewpoint corroborated by experimental data (Pianka 1978). However, similar patterns are present on the wings of nocturnal Saturniidae. They were also found in the fossil neuropterans Kalligrammatidae and Psychopsidae (Rohdendorf and Rasnitsyn 1980), which lived in the Jurassic, that is, before the pronounced radiation of avian fauna that began in the Cretaceous (Shimansky 1987) and, in any case, long before the origin of insect-eating Passerinae.

In general, each of the presented examples found in the lower taxa is normally explained by particular functions found in separate groups, but the explanation fails when we consider a broader spectrum of data on the morphology of neighboring groups. Consequently, functional interpretations are falsified by homoplasies and leave the structural traits unexplained.

The last example demonstrates true reasons for similarities originating at the lower taxonomic levels, avoiding the traditional functional interpretation. It is a unique case combining a survey of comparative morphology, physiology, and embryology, and dealing with the foot structure in neotropical salamanders. The web between the digits of these animals is normally considered to be an adaptation to an arboreal mode of life. It functions like a sticky vacuum sucker and provides definite advantages to salamanders that often walk on flat, smooth wet leaves of tropical trees or on cave walls (Alberch 1981; Wake 1991). Nevertheless, the web appears without any connection to life on vertical surfaces. Presence of the web correlates with fusion of the tarsal elements of the limb skeleton, and it originates during individual development if the number of cells in the foot vestige falls below a particular threshold. This can happen with a decrease in general body size (Alberch and Gale 1985; Oster et al. 1988; Wake 1991), which can be caused by various selective reasons. Although the functional importance of the web has been empirically and experimentally proven (Alberch 1981), this structure appears in evolution not as a reflection of the natural selection of climbing specimens, but as one of the possible outcomes of ontogeny, which are themselves determined by internal phenomena of the organism rather than external conditions of the environment. Thus, webbed feet should be considered to be preadapted (or proto-adapted, after Gans [1974]) for the life in trees and on walls used by some species, but not for all salamanders that possess this structure (Alberch 1981; Wake 1991).

The case of the salamander foot web shows that concepts of homology and homoplasy can be extremely close and even overlap. From one viewpoint, the web should be treated as a parallelism, because it is absent in the ancestral form and appears independently in descendants owing to selection by another trait, being a neutral feature itself. Moreover, one cannot determine that the web is not homologous in different salamander species, because identical developmental mechanisms occur in all species, since they are inherited from the common ancestor. Moreover, here we can see the difference between homology and homogeny, because the webs of different species are homologous, but not homogenous. This example also provides evidence that the reasons for morphologic peculiarities observed in species and genera are caused by morphogenetic factors instead. It is therefore unnecessary to continuously speculate on the functional significance of each structural feature of the lower taxa, unless strong evidence supports such hypotheses.

Summarizing this chapter, the following evolutionary perspective can be depicted. Similar morphologic characters independently appear in different species owing to shared morphogenetic mechanisms. Initially organisms do not use these morphologic novelties, since a coupled "structure-function" cannot appear prior to the origin of both the structure and the function. Therefore, any given structure originates as neutral relative to its future function (i.e., as a proto-adaptation, according to Gans [1988]), while the function is initially fulfilled by other structures. Only after the association structure-function has been established, can natural selection obtain the opportunity to fit the structure to the function somehow, changing the shape and size of the structure. It can be suggested that exactly this happened to the corrugation of the costal vein, secondary false veins, covering scales, and eyelike patterns on the wings of different insects. After these structures had originated, the first was adapted to protection of tracheae, blood vessels, and nerves in the wings of beetles, the second to better reinforcement of dipterous fly wings, the third to thermoinsulation in lepidopterans, and the fourth to warning coloration. Meanwhile, association between the structures and corresponding functions emerged only in some groups, whereas the structures remained unexploited in other insect taxa. Characteristically, the same feature can be used for different purposes in different groups. For example, the anal loop is supposed to facilitate coupling of the fore- and hindwings in particular groups and to play the role of an adaptation to semiaquatic life in others, whereas insects also exist that do not use this structure at all. In general, this situation accurately agrees with the general principles of animal functional and evolutionary morphology discussed in the appendix.

Form and Function:
A Review of General Concepts

What can lead to a way? Nothing but philosophy.
—Emperor Marcus Aurelius (A.D. 121–180)

In this book, I have attempted to show that some morphologic characters of insect wings can be interpreted as adaptations fulfilling certain functions. By contrast, there is a set of structural features that cannot be thought to be functionally significant. Future research might uncover adaptive reasons for the origin of these structures. However, this is not necessarily going to happen, because the features unexplainable in functional terms often find their explanation in terms of morphogenetic processes. This appendix can be used by those who are interested in further speculations on the subject: How can we differentiate characters that do play functional roles from those that are just artifacts of ontogenetic reconstructions? I will depart from the perspective of general functional morphology of organisms and try to show that functionally significant features generally characterize functional systems (e.g., the wing apparatus as a whole), whereas functionally neutral traits are mostly specific of single organs and their parts (e.g., wing venation).

Up to the present, many explanations of biological form involve a purely selectionist approach. Most research papers describe new structures, followed by hypotheses that suggest the way these structures could be used during the life of the organism. However, many biologists consider a minority of morphologic features to be adaptive (Mayr 1970; Schmalhausen 1983). The remaining properties are thought to be "by-products" of natural selection: thus they are a priori, left without any explanation. Meanwhile, it seems necessary to try to find opportunities for complete explanations in organismal morphology.

Basic Definitions

General biology is replete with terms that possess several widespread, non-coinciding interpretations. Discussion of the adaptive problem demands a preliminary definition of the main concepts.

A *taxon* is a group of organisms that is distinct enough to be given a name (Mayr 1970; Liubarsky 1993).

Evolution is defined as the process of transformation of taxa (Dullemeijer 1980), or as a change in morphofunctional organization (Margalef 1992), or as an inherited change of trajectory of normal ontogeny (Shishkin 1988), or as the "control of development by ecology" (Van Valen 1974). Hence, evolution consists in an inherited change of normal ontogeny of an organism, and that evolution leads to taxonomically significant reconstruction of the morphofunctional organization, resulting in a novel appearance relative to the preceding history of the organism.

Adaptedness is the ability of an organism to extract from the environment an amount of energy sufficient to maintain vital activity (Bock and von Wahlert 1965).

A *structure* is a spatial configuration assessed by the sense organs, primarily the eyes (Lima de Faria 1988). Structures can be reliably recognized among the other structures of an organism (Vedenov et al. 1972) and involve systems of organs and organs and their parts. Organs are more or less isolated parts of an organism (Schmalhausen 1938), to some extent separate and relatively independent (Liubarsky 1991b).

A *function* is an aspect of the interaction between the parts of an organism and between organisms and their environment (Vorobjova and Meyen 1988); this is perhaps its most general definition. In other words, a function is a flow of energy between structures (Lima de Faria 1988). Hence, function is often considered to be a change of the condition of a structure (Bock and von Wahlert 1965; Bliakher 1976; Dullemeijer 1980; Wainwright 1988; Lauder 1990; Iordansky 1994). By comparison, the definition of function as adaptation is more specific: function is the expedient reaction of living organisms, having an adaptive significance in interactions of the organisms with their environment (Matveev 1945, 1957; Kaganova 1972). As Kokshaysky (1980) stated, this is a definition of function at the organismal level. This particular interpretation of function is used subsequently. Structures are objects under study in morphology (Chebanov 1984), and functions are topics in physiology sensu lato (Schmidt-Nielsen 1979; Ugolev 1985).

Following Severtzov (1939), functions may be active and passive (hence structures can function actively and passively). A passive function involves changes in the condition of a structure, resulting from the action of an external (related to the structure) force. An active function involves force generated within the structure. Activity and passivity depend on the level of consideration. For example, the strokes of flying insects involve a passive function of wings and an active function of the pterothorax. On the molecular level all functions are passive; on the organismal level functions are mostly active. However, passive functions are also quite widespread (e.g., dispersal with wind and stream, warming in sunshine, perception of external signals, etc).

Functions and structures are described by *features* (also called *characters* or *traits*). Features are fragments of organization, delineated and named for a definite purpose (Liubarsky 1994; Bock and von Wahlert 1965): any property that demonstrates a similarity or difference in organisms, their parts or groups (Rasnitsyn and Dlussky 1988); a digit of morphologic or physiologic discretion of an organism (Lobashev 1963).

Faculties (after Bock and von Wahlert 1965) are pairs of features incorporating morphologic and physiologic aspects of organization, for instance, wing flight or eye vision. Faculties are idealized results of morphofunctional analysis and, hence, elementary objects in functional morphology. Faculties can be adaptive, neutral, or antiadaptive. If the latter, they reduce the difference between the normal (in comparison to related species deprived of this faculty) and necessary amount of consumed energy; thus, they diminish the probability that the organism will be preserved in future generations.

An *archetype* is an abstract model reflecting the construction of all the diverse forms included in a given group (Bliakher 1976; Kouzin 1987); it is a generalized plan of organismal design within a given taxon (Chaikovsky 1994). In other words, an archetype is a schematic diagram, a definite sum of connections between elements of the system (Svidersky 1962). *Elements* are presented by structures in the morphologic archetype (this particular definition has been undermined by classic morphology under the term *archetype*, after Owen) and by functions in the physiologic archetype (Vorobjova and Meyen 1988). Nevertheless, recognition of archetypes is a task for comparative functional morphology, because this procedure involves determination of interrelation and interdependence (i.e., functions) of parts of an organism. Accordingly, Liubarsky (1993) called the knowledge of functions a "glue" for morphology.

Archetypes are the central and most complicated concept of comparative morphology and have numerous conflicting interpretations. Often they are understood as ancestral morphology. This is an inadmissibly concrete definition. A generalized ancestor can occupy even a more distant position from the archetype than recent groups, each having found its own way of deciphering the schematic diagram, each having understood and developed the main idea of the scheme in a particular way. An ancestor has just entered the adaptive zone of the taxon; it is not abundant and fits the new environment poorly, because its parts are not yet well co-adapted to fulfill the peculiar functions demanded by the new conditions of life. Evolving descendants must form a new adaptive zone, using primary opportunities to greater or lesser extent and adjusting interactions (functions) between particular structures. Thus, an archetype (a generalized morphologic description of a taxon) arises in parallel with an adaptive zone (a generalized ecologic description of the taxon) and is not something given in advance.

An original concept for understanding the archetype was suggested by Beloussov (1980). He concentrated on the fact that the ontogeny of different classes of vertebrates shows not only divergence, but initially shows convergence of individual development towards a "knot of similarity," which in the Vertebrata is presented by the stage of axial organs that has the same morphology in all the classes. Accordingly, Beloussov stated that Owen's archetype of a taxon should be depicted not as an adult form, but as an embryonic stage, corresponding to the knot of similarity of the taxon.

Blocks of the schematic diagram are called *merones* (after the Greek μεροσ = part) (Meyen 1978; Shreider 1983; Chaikovsky 1990; Liubarsky 1993); each merone has its own name. Correspondence of parts of real organisms to merones of the archetype constitutes Owen's "general homology." Thus, the procedure of homologization does not demand recognition of a common ancestor. Nevertheless, once the ancestor has been recognized, its parts can be brought into agreement with the archetype, that is, homologized in a general sense. Having been homologized, the parts of the ancestral organism can be named.

Owen's concept of special homology (or Lankester's homogeny) is applicable to structures found in different organisms, if these organisms can be thought of as descending from a common ancestor, and whenever these structures correspond to the same part of the ancestor's body. Different structures in different beings (recent and/or extinct, adult and/or embryo) are treated as homologous, if they are characterized by similar location and/or

TABLE A1

Correlation of concepts, designating different types of secondary similarity in organisms

	Homoplasies	
	Adaptive	Nonadaptive
In homologous structures	Homoiology	Parallelism
In nonhomologous structures	Convergence	Isomorphism

construction. Many "and/or's" are because any of the criteria can be violated. A situation is imaginable that meets no single criterion, since evolution changes the construction of parts, their allocation inside the organism, the way of generation in ontogeny and other special details. However, merones generally remain recognizable. "Special homology" is a completely secondary concept in relation to "general homology," because in actual practice the ancestor is recognized by its similarity to the archetype, which has already been derived based on comparative study of descendants.

The *morphospace* of a taxon is the sum (or better—integral) of the archetypes of all the subordinate taxa. It can be described as a hypervolume in hyperspace of features of the merones of all the organisms included in the taxon (Meyen 1978)—a unity of features embracing all the diversity of the taxon (Chaikovsky 1994; see also Liubarsky 1991a).

Two or more different organisms are called *similar* if they share common features. Characters of primary similarity are inherited from common ancestors and are synonymous for homogeny (or special homology). Characters of secondary similarity, which are absent in the nearest common ancestor and originate later, being independent in distinct pathways of evolution, are called *homoplasies* (Russell 1916; Futuyma 1986; Sanderson and Donoghue 1989; Wake 1991; Rasnitsyn 1986, 1992a; Beklemishev 1994). As we have already seen, primary and secondary similarities are not strictly distinguishable concepts (see sec. 4.9). It is also impossible to state a strict distinction between different types of secondary similarity. However, homoplasies can be conventionally distributed among four categories (tab. A1).

Parallelism constitutes nonadaptive secondary similarity of homologous parts (Timofeeff-Ressovsky et al. 1977; Severtzov 1990). If organisms share a sufficient number of merones, they are placed in related taxa; if parallelisms are observed simultaneously in several merones, parallel evolution of the taxa must be taking place (Bigelow 1958; Schmalhausen 1969; Raup and Stanley 1971; Meyen 1973; Berg 1977; Menner and Makridin 1988; Iordansky 1994).

Convergence (or analogy after Aristotle) is often observed in equivalent environments. It constitutes adaptive homoplasy of nonhomologous structures. Different merones are obtained by nonrelated beings; hence, convergence is thought to be an attribute of the evolution of phyla that diverged long ago (Darwin 1872; Schmalhausen 1969, 1983; Bliakher 1976; Pianka 1978; Mednikov 1980; Simpson 1980; Vorobjova 1980; Dogel' 1981; Severtzov 1987; Menner and Makridin 1988; Iordansky 1994).

Homoiology after Plate constitutes adaptive homoplasy of homologous organs, say, the parietal crest in gorillas and hyenas or the breastbone in birds and moles (Bliakher 1976).

A logical possibility for the fourth category exists, namely nonadaptive secondary similarity of nonhomologous parts. Such similarity has been described in plants for flowers and inflorescences (Meyen 1974). Examples are not numerous (this can be a temporal situation, owing to low anticipation on the basis of up-to-date biologic concepts) and have no special term. They might be designated as *isomorphisms* (Bliakher 1976). Presumably, isomorphisms are caused by developmental patterns that have not been inherited from a common ancestor, but are at the same time similar. This can be anticipated when developmental processes are comparatively simple and based on a small number of logical opportunities, so that secondary similarity is acquired regularly, but in fact on a random basis, owing to purely probabilistic reasons and without a common factor such as similar demands of environment.

Complementarity of Form-Generating Factors and Uncertainty of Explanations in Morphology: Recognition of the Problem

Modern biology states that the evolution of an organismal morphology is simultaneously influenced by two factors: natural selection (external, in reference to the organism), and embryonic processes that determine allocation of differentiated tissues in ontogeny (internal factor). Both factors only partially depend on previous evolution of the organism (see further explanations below).

The action of natural selection is evident from observation of analogies, which are convergent features acquired by beings of different organization under adaptation to similar environmental conditions. Canonical examples of analogies include the streamlined body shape of large fast-swimming vertebrates; the robust extremities in digging animals; the characteristic profile of wings, fins, and flippers (thickness always decreasing from the leading to the trailing edge); and the sphericity of some species of Cactaceae and Euphorbiaceae.

Rigorously stated, evolutionary convergence does not strictly demand possession of one particular function. If quite different functions exploit a common physical principle, that is enough (stated by G. B. Kofman, personal communication). A priori the number of such principles and of organic constructions built around them is not large, and nature may not be able to use them all. So a particular task that natural selection imposes on a changing organism can be resolved in a limited number of ways. This explains a long range of convergences in functional systems in very different organisms: animal wings and flying seeds of plants (Ennos 1989d); eyes in vertebrate and invertebrate animals (Alexander 1988); thermal insulating coverings in mammals, birds, and insects; swim bladders in aquatic animals and plants; acoustic and vocalization organs in diverse animals.

The most general reason for the origin of convergent similarity seems to be minimization of substance and energy cost: regular shape is explained by economy of material (Mordukhaj-Boltovskoj 1936). Hence, as only few symmetry types exist in three-dimensional space, organisms resulting from the selection for the minimal amount of necessary matter inevitably fall into several groups of similarity defined by their type of symmetry. The same can be said about forms, such as the hexagonal facets in insect compound eyes, honeycombs, or epithelial cells. These shapes evolve because hexagons are geometrical figures that fill surfaces with the least perimeter of cell walls. In a similar way, icosahedrons are regular ridge polytopes characterized by minimal surface area for a given volume; this determines the morphology of some viruses and radiolaria. Also, minimization of surface area to body volume ratio leads to sphericity, irrespective of the role this minimum plays in the life of the organism. Examples include the deceleration of evaporation in arid zone plants, the decrease in the probability of meeting enemies in colonial protozoans, diminishing heat loss, or something else. Exploitation of common physical principles explains not only analogies in organismal construction but also those in inert and ideal objects (Stevens 1973; Petukhov 1981; Vasiutinsky 1990; Zimov 1993).

The influence of morphogenetic factors is primarily observable in the parallel evolution of related forms. The best elaborated examples are given by the secondary similarity of anuran legs (Alberch 1981; Alberch and Gale 1985; Oster et al. 1988; Wake 1991) and shells of *Cerion* (Gould 1984, 1989). In general, it can be considered to be proved that morphologic variability is not accidental, but instead results from the action of particular morphogenetic mechanisms, which are able to create only a limited number of patterns; evidence is given by vast data of comparative embryology and teratology (Krenke 1927; Seilacher 1974; Beloussov 1975; Lande 1978; Maynard Smith 1978; Alberch 1980; Gould 1984, 1989; Oster et al. 1988; Belintsev 1991; Saveljev 1993).

Related species share common morphogenetic mechanisms. Each mechanism can produce only a limited range of morphologic variants of a structure. As a result, organisms generally are able to engender only a restricted spectrum of different descendants. The latter is expressed in the phenomenon of epigenetic landscapes (Waddington 1940). The more related two taxa are, the more complete is the coincidence of their morphologic diversity (Vavilov 1968). Practically speaking, Vavilov's homologous series in the variability of several taxa suggest that those taxa possess common developmental mechanism(s) of pattern formation; thus they are related. A range of modalities of a particular feature occurring in many species defines (together with the mechanism of their generation) a feature of a high taxon (genus or family). The latter consideration is used in reconstructions of phylogenies (Chernykh 1986).

If an organism and the processes of its formation are simple enough, comprising a comparatively narrow range of logical opportunities for construction of each structure, purely probabilistic reasons suggest that homoplasies will appear regularly in quite distant groups (families and orders). This is more apparent and common in plants than in animals (see the long list of examples given by Went [1971] and Meyen [1973]). Bacteria also display secondary similarity owing to their simple morphology; this imposes strong handicaps for phylogenetic reconstructions (Zavarzin 1969, 1974).

Morphogenetic factors are independent of natural selection within evolutionary sequences; the range of forms an organism can engender is totally determined by its epigenetic landscape and preexists in reference to the beginning of an elementary evolutionary change. Based on this consideration, the structuralist paradigm suggests a triple approach to the explanation of biologic form, using physiological, morphogenetic, and historical arguments (fig. A1A) (Seilacher 1974; Thomas 1978, 1979). From my viewpoint, the historical factor adds unnecessary complication in this scheme. It would be simpler and quite adequate to include time in the morphofunctional system,

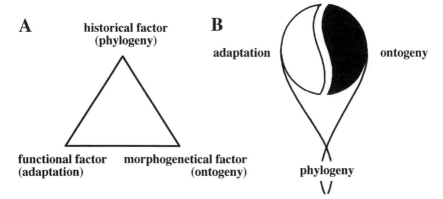

Figure A1. General approaches to the explanation of form in organisms: (*A*) structuralist approach; (*B*) approach suggested hereinafter.

since history reflects the collaborative action of previous morphogenetic and selective reasons (fig. A1, *B*).

Mutual independence of the two pattern-forming factors leads to the necessity of bringing to biology Bohr's concept of complementarity, according to which biologic structures have a dual nature. Bohr himself repeatedly wrote about the ubiquity of typical complementarity in biology (Bohr 1971, pp. 398, 490, 524). This suggestion was later supported by numerous authors (Meyen 1973, 1974, 1975; Korochkin 1986; Gould 1984, 1992; Arnold et al. 1989; Chaikovsky 1990, 1994; Belintsev 1991; Berg 1993; Roth et al. 1993; Thomas and Reif 1993; Vakhrushev and Rautian 1993; Koucherov 1995). Nowadays it is unclear (Liubarsky 1993) whether the classic interpretation of Bohr's term as it is used in quantum mechanics can be used in biology. Physical science deals with micro world phenomena, which have nothing in common with things in the macro world and thus cannot be described in equivalent terms (Heisenberg 1969). Otherwise, complementarity in physics means "neither this, nor that," whereas in biology "both this and that." In any case, the need for this concept arises in studying phenomena that appear in two distinct contexts; whenever there exists a specialized language for description of each aspect, but the languages are not mutually translatable (G. B. Kofman, personal communication; similar in Korochkin 1986). The concept of complementarity indeed admits that two factors are acting and that we do not know how to combine them within a single framework, nor how to reduce the biologic complexity into simple, mutually correlated, and at the same time adequate terms. Complementarity is kept in use, until a special language has been elaborated for description of the phenomenon under study. A similar situation exists also in psychology, history, economics, and other sciences, which Scriven (1959) called *irregular subjects*.

Based on the complementarity concept, biologic diversity at first approximation should be considered as an intersection of two sets: "forms able to appear" and "forms able to survive," resulting in the subset of "forms that have appeared and survived" (Raup and Michelson 1965; Seilacher 1974; Thomas 1979; Gould 1984, 1989; Thomas and Reif 1993). Both sets are a priori much larger than the intersection. On the one hand, the widest opportunities of pattern formation in any species are evident from the vast number of races and breeds of domestic animals and cultivated plants, the external difference among which is much larger than that among wild species. On the other hand, wheels and propellers represent highly effective locomotor organs, but no such systems occur in organisms. As a consequence, a complete explanation of organic form neither can be "purely morphological" (after Geoffroy), nor can it be based on the sole knowledge of functions (after Cuvier).

Mutual redundancy and independence of the sets have engendered the concept of constraint, which is the source of restrictions on change that do not arise through the action of stated causes within a taken theoretic frame (Gould 1989). As a result, each of the two pattern-generating factors obtains its own "scope of competence" with specific rules for deduction. In fact, necessity is determined by one group of rules and opportunity by another. The two scopes are divided by space, which is impenetrable for logics. Therefore, in theoretical morphology, like in quantum mechanics, the dual nature of structures and complementarity of factors results in uncertainty of any deduced statement. An imaginable deterministic prognosis can be built around either selective or morphogenetic considerations, and there are no rules for logical transition between the two groups of phenomena, which are totally random in relation to each other.

Uncertainty represents the main theoretical problem treated in this part of the book. I will try to show that the features of an organism can be approximately divided into two categories, one owing to natural selection and the other mainly to morphogenetic regularities.

The following account demands consideration of two basic concepts of functional morphology, which seem to need no special corroboration and have no exceptions in relation to the description of living organisms. These basic concepts are hierarchy of faculties and multivariate correspondence

A) Hierarchy

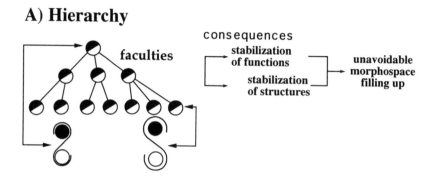

B) Multivariate correspondence

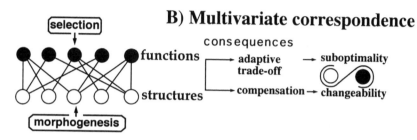

Figure A2. The correspondence between main general concepts of evolutionary and functional morphology. For terms and explanations see text.

between structures and functions (fig. A2). "Consequences" derived from principles are not corollaries in the mathematical sense. Owing to ubiquitous complementarity, hardly a statement in biology univalently follows a more general concept. In biology consequence is a notion based on a commonly accepted idea, but usually with an additional meaning in relation to the idea.

Hierarchy of Faculties

A hierarchy is a series in which each member is included in the member that precedes it and includes all those members that follow it (Medawar and Medawar 1983). Hiearchy is an aspect of biologic complexity, which was broadly discussed in connection with the concept of structural and functional levels in living systems (Bykhovskii et al. 1972; Timofeeff-Ressovsky et al. 1977; Riedl 1977).

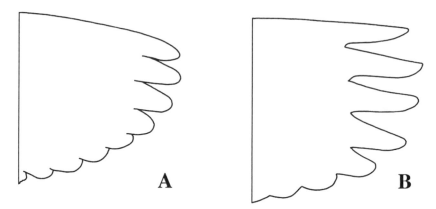

Figure A3. Distal parts of the wings of the squacco heron *Ardeola ralloides* (A) and of the common heron *Ardea cinerea* (B). (After Kokshaysky 1988a, 1988b.)

The Principle of Evolutionary Stabilization of Functions

All structures and functions of organisms are constructed hierarchically (fig. A2*A*). This is directly connected with the principle of evolutionary stabilization of functions proposed by Kokshaysky (1973, 1980, 1988a, 1988b). According to this principle and to accommodate changed conditions, minor alterations are made in peripheral details of the corresponding form-function complex with minimum change of the function itself. This also occurs in a more general formulation: during evolution any system retains "systemic" similarity at the expense of deflections from geometric and other particular types of similarity (Sahal 1976).

Kokshaysky illustrated the principle of evolutionary stabilization with several cases of the dependence of morphology on the absolute size of specimens. Two cases are presented subsequently. The first example deals with wings and flight in herons, an ecologically uniform family of birds.

The planform and flight kinematics of wings are similar in all heron species, the only difference being that the species-specific number of slots between marginal feathers constantly increases with body size (fig. A3). Slots manage the angle of attack and prevent separation of airflow (for this concept, see sec 2.2) from the wing edge, because this separation becomes more dangerous with the increased size of the bird.

The general interpretation of this simple fact is that in the size range of herons, an increase in absolute size is accompanied by small but regular violations of geometric and dynamic similarity in the distal zone of the wings.

In the long run, these minor alterations favor preservation of geometric, dynamic, and physiologic aspects of similarity of flight in this family. In other words, the described regularity assures that herons of different size (the biggest are more than twelvefold the weight of the smallest) share general features of the structure, biomechanics, and physiology of their flight system, keeping a constant ecologic similarity (or adaptive zone) within the family.

Other bird families overcome the difficulty raised by the increase in total mass in different ways. One way involves intensification of flight accompanied by narrowing and/or shortening of the wings, more frequent wing beats, greater velocity, and smaller angles of attack. However, this would change the appearance and ecologic niche typifying herons into one typical of ibises. An opposite possibility consists in reducing the angle of attack, widening and inactivating the wings, and adopting the soaring flight typical of storks. Only the emargination of the wing-tip feathers enables large herons to keep a heron morphology and to continue to occupy the ecologic niche characteristic of this family, involving hunting in wait for prey and assessment of feeding territories from aloft (Kokshaysky 1988a, 1988b).

Kokshaysky's second example considers the scaling of locomotion of quadrupedal mammals. Galileo was the first to notice that as animals increase in size, they must have an allometrically more robust skeleton to resist gravity. The proportions found in nature have been explained using the concept of elastic similarity, a hypothesis corroborated by data on limb bones of adult ungulates with excellent agreement between data and theoretical model (McMahon 1973, 1975). Nevertheless, a consideration of material on a wider range of mammals from seven different orders showed the preservation of simple geometric similarity (Alexander et al. 1979). This is unclear from the traditional viewpoint, especially since the bones of small and large mammals possess identical strength and tension indices (Biewener 1982, 1983). The entire range of mammals could display a mechanism that overcomes the increase of locomotory loads on bones, and this mechanism could be dynamic rather than static. Indeed, along with the size increase in mammals their legs become progressively straightened (fig. A4). This reduces the moment of the force resulting from interaction with the ground.

So, what is stabilized in the second example? Perhaps in the "ungulate" case this is the group-specific type of locomotion associated with peculiar skeletal construction and arrangement of muscles (elongated distal elements of legs, leg operating as a simple lever, moving only back and forth). The stabilization allows general kinds of similarity (including ecology) and bene-

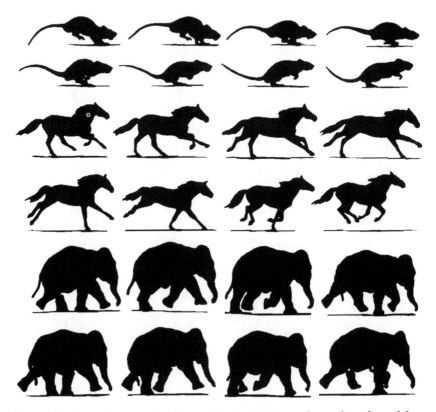

Figure A4. The tendency to straighten limbs in supporting phase of quadrupedal locomotion in mammals under an increase in size. (After Gambarian 1972, from Kokshaysky 1988a, 1988b.)

fits from the elastic similarity. Ungulates form a morphologically and ecologically uniform group of animals, so differentiation in size is an important mechanism of their diversification. Furthermore, the choice of methods usable by larger animals for resisting gravity is restricted. Extreme changes in functional morphology can make the species lose adequacy to the adaptive zone; hence a passive increase in bone thickness is practically the only possible way in which ungulates can resist increased loading. Quadrupedal mammals do not maintain a particular locomotory type, but rather the very opportunity to run. This is permitted through a huge range of mass values, at the expense of abrupt changes in morphology and biomechanics. The general geometric similarity of mammals is quite evident and documents the disadvantages of extra thickening or thinning of the bones (Kokshaysky 1988a, 1988b).

In general, treatment of data on mass-dependence of different morpho-functional characteristics by regression analysis requires the consideration that the initial data can be approximated to a line that describes the common tendency in the entire group. At the same time, particular clouds of dots also exist that are characterized by "local" regression lines to mass. These sub-ordinate lines indicate morphologic, functional, and ecologic peculiarities of subordinate taxa. Particular and general lines can be quite distinct and have different slopes. First described by Kokshaysky (1970), these subsets were later explored in greater detail, as reviewed by Schmidt-Nielsen (1984), who designated them as "secondary signals." Examples have been adduced for mass-specific data on the metabolic rate in mammals, circadian energy budget, and egg weight in birds.

Another example of an adaptive radiation pattern that matches Kokshaysky's theory well is cranial construction in amphisbaenians (Gans 1974). Amphisbaenians comprise a sister group to lizards (or lizards + snakes) and occupy a subterranean mode of life, burrowing through the soil with the help of their cranially reinforced heads. This burrowing function is displayed in several modifications, and the skulls of the amphisbaenians adapted to each modification differ so markedly (fig. A5) that it is very difficult to believe that these animals are related. Nevertheless, all these profound changes affect only the visceral part of the skull, whereas the cerebral, brain-containing section remains equivalent in all species, independent of the mode of burrowing used.

Kokshaysky's ideas are strongly supported by the extensive material developed by Shwartz and his followers (Shwartz et al. 1968a, 1968b; Shwartz 1980), representing the Sverdlovsk (Ekaterinburg) school of Russian/Soviet ecologists that began in the 1950s. Their results seem to be unique in the world literature. They investigated adaptations to extreme environments at the levels of species and populations. Shwartz's work showed that whenever related species inhabit contrasting environments (north-south, lowland-mountains), they do not differ significantly in the major morphofunctional characteristics (relative weights of brain, heart, lungs, liver, pancreas, bowels and kidneys, and hemoglobin concentration). However, different populations of particular species, inhabiting an equivalent range with similar contrasting conditions, differed greatly in these indices. For example, the montane populations of a normal lowland species were characterized by increased heart volume, whereas nearby but typically montane species had hearts of the usual size. This result seemed paradoxical, since populations exhibited obvi-

Figure A5. Skull shapes in amphisbaenians: (A) *Amphisbaena bakeri;* (B) *Ancylocranium ionidesi;* (C) *Monopeltis jugularis.* (After Gans 1974.)

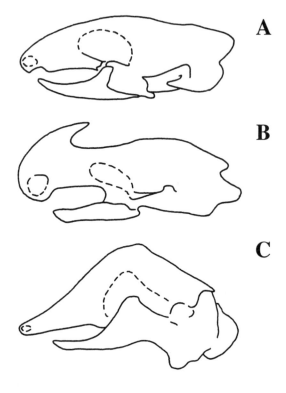

ous adaptations to extreme environments, whereas species did not. Nevertheless, the result was corroborated by outstanding data for several hundred species including all major groups of terrestrial tetrapods.

Shwartz suggested that changes in organs functionally connected with diverse features of the organism are unlikely to be energetically advantageous, because such changes disturb a historically achieved morphofunctional coordination. According to Shwartz, macromorphologic deviations are less effective than those at the level of molecules, cells, and tissues (i.e., changes at lower levels of organization), for instance by different variants of enzymes, structural proteins, and intracell ionic media. These changes tend to result in tissue incompatibility and thus in reproductive isolation, which is therefore the consequence of speciation rather than a cause. Consequently, new species will differ in minor features, rather than in the features of the generic archetype. The latter can be violated and changed in the beginning of an evolutionary change, but it is thereafter rebuilt under the form-generating influence of stabilizing natural selection.

The arguments of Kokshaysky and Shwartz differ mainly on the rank of considered taxa; thus Shwartz discusses adaptations in populations and species, whereas Kokshaysky examines families and orders. Several similar examples are known (LaBarbera 1990; Nikolajchuk et al. 1991) with practically identical conclusions in all cases: the generation of a new taxon commonly leaves the main features of organization (archetype of the upper rank) unchanged. Adaptation to a new environment commonly involves change of parts that are less important for the organism as a whole.

Another Formulation: The Principle of Minimum Change

In all likelihood, the principle of evolutionary stabilization of functions results from the elements of high hierarchical levels interacting with a larger number of other parts of the organism. Correspondingly, the higher the level of morphofunctional change the more exclusive the reconstruction of the organism. Meanwhile, evolution proceeds by a path along which the total sum of changes is minimal; evolution follows the least change of organization. Hence, evolution is directed, at least on intervals between bifurcations.

Earlier in this century, Rashevsky (1943a, 1943b, 1954, 1965) introduced the principle of maximal simplicity in organismal morphology. He stated that a particular structure or a construction found in nature is the simplest among possible structures and constructions able to fulfill a given function. However this principle can hardly be used in this form. For verification, it needs formalization and illustration with particular biologic examples. Nevertheless, it is unclear how to define the maximum value of simplicity.

By contrast, usage of the principle of least change submits the problem to mathematic investigation. Instead of determining the absolute simplicity of a construction, only the magnitude of its change relative to the previous condition must be measured. This is much simpler to do, because describing a complex figure is a much more difficult (and almost impossible) task than comparing two complex figures (D'Arcy Thompson 1942).

In fact, Kokshaysky provided a biologic reformulation of the Maupertuis (1698–1759) physical principle of least action. The higher the level under consideration, the less likely the changes to it; structures and functions tend to be very conservative instead (Mayr 1970; Lauder 1990; Vorobjova 1991). Such stability does not mean absence of variation, but instead just the opposite. Variability is the basis of each new clade, but it is given a direction from previously existing structure (Schmalhausen 1945), since "the type rules all its varied modifications" (Russell 1916).

The principle of stabilization of functions used in evolutionary physiology is the parallel of the principle of evolutionary stabilization of structures (or the unity of type) in morphology. Both rules together indicate action of the principle of the minimal change on functions and structures correspondingly. This explains the generally inert nature of evolution. The idea shows why the number of taxa decreases with the growth of rank, that is, why new classes originate more rarely than new species.

Unavoidable Filling of Morphospace

The principle of minimum change has the consequence that corresponds to the course and rate of evolution. A taxon will not become another taxon until all the constructions permitted by the morphology of the lower taxa have appeared and been tested by natural selection. Mamkaev (1983, 1984) seems to have been the first to suggest a similar statement, namely that any system of an organism can appear in a restricted number of morphofunctional variants, and that all these variants will appear in evolution. An illustration of this statement constitutes the pattern of evolutionary diversification of the main types of insect flight kinematics provided in chapter 3. Another example is given by the evolution of the projectile tongue in salamanders (see references in Wake [1996]): for the tongue projection, skeleton modification is necessary. Only two arrangements of skeletal elements are possible, and they both have been used within different lineages of salamanders, giving the same functional result.

The principle of the necessary filling of morphospace is analogous to the movement of liquid over a surface: the farthest edge of the flow only progresses after any cavity it meets on its way has been filled. Such a sequence of events can be compared to the main trends in evolution, which were proposed by Severtzov in his classic survey on evolutionary animal morphology (Severtzov 1939) (fig. A6). His scheme includes two main directions called *arogenesis* and *allogenesis*. The first involves the progressive change of an organism, making it more complex (more differentiated) and thereby able to survive under a broader range of environmental conditions. The morphofunctional result of arogenic evolution is acquisition of a set of novel features enabling the organisms to enter a new adaptive zone. This set has been named *aromorphosis* in Russian literature (after Severtzov 1939); in the West a more common term is *adaptive syndrome* (Gans 1974). Simple examples of adaptive character sets are given by differentiation of blood circulation into two cir-

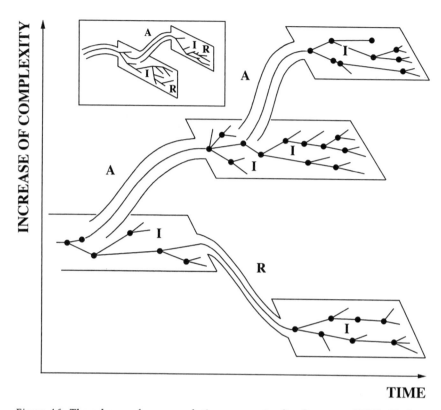

Figure A6. The scheme of macroevolutionary events after Severtzov (1939). Circles correspond to species, planes to higher taxa Meyen's archetypes, or potential morphospaces. Severtzov recognized three main types of morphologic change: "arogenesis" (*A*), "idioadaptation" (*I*), or allogenesis, and regress (*R*). The latter can be thought of as a type of allogenesis, because any adaptation to a particular narrow environment is followed by simplification of organismal construction (see text). To combine the concepts of Severtzov's idioadaptation and regress, only inclination of the scheme is necessary, as in the fragment. Fitting well the neo-Darwinian paradigm, this scheme shows only divergent processes, while convergence is at least as common an evolutionary phenomenon. (Berg 1977.)

cuits in terrestrial tetrapods, segregation of the digestive system in mammals and insects, endothermy, and differentiation of brain and sense organs. The origin of insect flight (and correspondingly the origin of pterygote insects, comprising the absolute majority of species) is a typical case of a novel adaptive syndrome.

Allogenesis is the adaptation of an organism to a particular environment through acquisition of minor morphofunctional peculiarities. Moreover, arogenesis and allogenesis are changes on correspondingly high and low levels of organization. Thus, owing to the principle of the necessary filling of morphospace, evolution continues its allogenic ways until all the principally permitted states are tested. Only then will a new adaptive syndrome evolve and many organismal features change. For example, acquisition of feathers by bird ancestors permitted the evolution of complex thermoregulatory mechanisms and enabled flapping flight, which caused aspects such as pneumatization of bones, differentiation of feathers, differential increase in muscle mass, typical breastbone and air sacs, shortening and fusion of caudal vertebrae, complex brain and behavior repertoires (including navigation), weakened sense of smell, and strong vision and hearing (Rautian 1988).

Appearance of a new type of biologic organization will inevitably cause eventual reconstruction of ecosystems and, hence, subsequent evolution of organisms that inhabited those ecosystems earlier. Therefore, emergence of a new level of organismic complexity is followed by numerous changes at lower organizational levels. As a result, the phylogenetic tree remains dynamic: on the geologic scale taxa constantly appear, disappear, and reemerge, like the branches of a tree that constantly grow longer, emerge, and disappear, with the caveat that thicker branches are more stable and will not reemerge once they have disappeared.

The principle of the unavoidable filling of morphospace illustrates why groups that undergo evolutionary morphologic simplification are more numerous than those that are selected for increased complexity. Reductional mutations are very abundant (Serebrovsky 1973; Schmalhausen 1983), and reduction can be more simple morphogenetically, whenever the structure to be reduced lies on the periphery of embryonic organizers (for this concept see Waddington [1940], Saxen and Toivonen [1962], Gans [1974], and Beloussov [1987]). Accordingly, ecologic specialization tends to be followed by relative simplification of the construction. In taxonomic analysis, features describing reduced organs are lighter and less important than progressive (more complex) changes (Liubarsky 1991b).

The necessity (or relative feasibility) of using all available constructions permits a group to decelerate a generally progressive succession of evolutionary changes. This is evident on the geologic scale and has been depicted in Severtzov's (1939) scheme (fig. A6).

The principle of necessity shows the perspective of evolutionary prognoses, that is, our ability to more or less definitely foresee in what manner organisms are going to change after another significant change of environment (the latter can be a result, e.g., of human impact on the biosphere). The main problem that must be resolved is a procedure for the analysis of morphospace. In Severtzov's scheme, this is symbolized by a plane and acts as the "cavity" on the way of evolutionary change.

An essential property of the morphospace of a mature taxon is that it is structured; any morphospace performs a particular set of morphologic conditions. This is caused by the interaction of two facts. First, any feature varies within definite limits. Hence, all species of a taxon are likely to lie within a limited volume of n-dimensional space. Second, taxon evolution is inevitably followed by adaptive radiation (branching of phyletic lines), which is connected with specialization of subtaxa in the use of specific kinds of resources, whereas the forms intermediate between specialized subtaxa become extinct (Giller 1984; Chernykh 1986; Iordansky 1994; Rasnitsyn 1996). In other words, maturation of a taxon is the process of splitting the primary continuous opportunity space into a number of nonoverlapping cells.

However, divergence of subtaxa and partitioning of an environment (formation of new "cells") cannot proceed infinitely, since the resources in any ecosystem are limited. Furthermore, the number of specimens comprising a viable population cannot fall below a certain value (Soule 1987), and supporting vital activity of the minimum number of specimens demands energy. Consequently, each "cell" of an environment must be able to supply its inhabitant population with at least a minimum of energy. This corresponds to the early idea by Timofeeff-Ressovsky (1995) that morphologic diversity can be described as a set of quanta, or units, further indivisible at the acquired level of taxonomic analysis.

Several previous descriptions of the space of morphologic opportunities deal with rather simple shapes: mollusk shells (Raup and Michelson 1965; Raup 1966, 1967), bacteria (Zavarzin 1974), and hydroid polyps (Beloussov 1975). More complicated cases are considerably more problematic (Gould 1984; Thomas and Reif 1993). For example, a profound survey of butterfly color patterns conducted by Nijhout (1991) showed that morphospace is limited by more than three variables; thus, its representation is quite sophisticated and can be carried out at a generic level rather than a higher one. Above all, Raup and coworkers documented potential and real morphospaces; the second is a part of the first as an evidence for the fact that not all ontogeny-permitted constructions are adaptive (Raup and Stanley 1971). Nijhout (1991) developed these ideas further, having pointed out nondirect correspondence between developmental space and morphospace: More than one point of developmental space can correspond to a point in morphospace; but morphospace also contains points that have no prototypes in developmental space; such morphologic conditions would be prohibited ontogenetically, because they cannot be achieved by means of any adjustment of developmental parameters.

A general procedure for morphospace quantification was proposed by Meyen (1973, 1974, 1990) as the concept of repeated polymorphic sets (RPSs). Each RPS constitutes a series of morphologic states (modalities, according to Rasnitsyn and Dlussky [1988]) of a given merone (Chebanov 1977, 1984; Chaikovsky 1990). The principal property of a RPS is its recurrence. Any member of an RPS can be derived of a neighboring modality by a transformation that is the same for all the series. To be exact, an RPS is D'Arcy Thompson's (1942) transformation trajectory divided into standard (or similar, in mathematical sense) segments.

Meyen's concept promises so much to theoretical biology that the most surprising aspect of the recurrent polymorphic sets is that they do indeed exist. Their reality was first shown by Krenke (1927) in his outstanding work on the shape of plant leaves. In addition to the particular RPS given by Meyen and Krenke in their cited articles (data on leaf segmentation, morphology of spores and pollen, epidermal structures, vein anastomoses, and plant teratologies), several analogous examples developed from the famous publication by Hutchinson (1959), reviewed by Maiorana (1978) and Giller (1984): sympatric species of a genus form a series with their body and/or mouthpart size with neighbors differing by a factor of 1.28. These examples illustrate the result of competition in splitting initially undivided morphospace representing a single genus into cells corresponding to species. Likewise, each of Hutchinson's series is characterized by a specific kind of similarity, and the process of speciation that led to the origin of recurrent series is evidently analogous to that given by the examples of Kokshaysky (1988a, 1988b).

As with all regularities in biology, the Hutchinson-Meyen rule has its exceptions. Nonetheless, it illustrates well opportunities to calculate morphospace and shows perspectives of predictive biology. In any case, any science starts from what is present, but sooner or later begins to consider things that are not evident but that could also exist (Kolmogorov 1991).

Multivariate Correspondence between Structures and Functions

Correspondence between morphologic and functional archetypes is incomplete (Vorobjova and Meyen 1988), because form and function are not directly connected (Iljichev 1973; Malakhov 1980). This conclusion was given concrete expression by Maslov (1980, 1984), who put forward the principle of multivariate correspondence between structures and functions: any organ fulfills more than one function, and each function is supported by multiple morphologic structures (fig. A2, B). This phenomenon was also discussed by Severtzov (1939), Kokshaysky (1980), and Fabri (1980).

Superposition of hierarchical (fig. A2, A) and multivariate (fig. A2, B) schemes gives the schematic diagram of biologic organization. Within this diagram one can model different rules of structural and functional evolution (for examples see Dohrn [1875], Russell [1916], and Plate [1928]). These rules were first systematized by Severtzov (1939) and intensively discussed and strongly supplemented after his basic survey (Matveev 1945, 1957; Vorontzov 1961, 1967, 1989; Arshavsky 1968; Kokshaysky 1980, 1988a, 1988b; Shwartz 1980; Matienko 1981; Schmidt-Nielsen 1984; Ugolev 1985; Alexander 1988; Natochin 1988a, 1988b; Mamkaev 1991; Swartz and Biewener 1992; Tystchenko 1992; Thomas and Reif 1993). Following Severtzov (1939), these rules, however, should be considered as particular cases of evolution rather than stable general biologic regularities, because a single evolutionary change of organism can be accompanied by nearly all kinds of structural and functional change simultaneously.

The principle of multivariate correspondence has two consequences, which are of fundamental importance for functional morphology.

Consequence 1: Adaptive Trade-off

Organismal construction fits the principle of adaptive trade-off (Norris and Lowe 1964; Gans 1983, 1988; Rasnitsyn 1984, 1987; Dudley and Gans 1991): each structure functions in a manner that does not interfere with the functioning of other parts of an organism. Consequently, all structures tend to be more or less functionally suboptimal, because each function establishes its own conditions on the construction of an organism. Because conflicting functions waste energy, natural selection acts to decrease energy expenditure of the organism and to achieve a compromise among all functions. Accordingly, the organism as a whole set of faculties corresponds to one of the local minima of the space of constructional opportunities. In other words, the overall organism (but not its parts) is locally optimal.

The principle of adaptive trade-off has its own consequences (fig. A2*B*). First, any change of organismal construction can be achieved only via an inadaptive state. Inadaptation is a one-sided change of few features, not followed by corresponding changes in the general construction of the organism (Rasnitsyn 1986). Thus, inadapted beings are characterized by a broken trade-off condition, are evolutionarily nonstable, and are normally displaced from their niche by more adaptive (eu-adaptive) forms that can persist for long periods of time owing to their better-balanced trade-off construction. Paraphrasing the well-known saying by Churchill, trade-off is awkward, but everything else is worse. Because of this trade-off, evolution is not monotone: transition among relatively stable states proceeds via instable states, which correspond to inadaptive phases of the evolutionary cycle (Rasnitsyn 1986; Kreslavsky 1991). We may therefore assume that evolution is possible only in "ecologically soft" conditions lacking strong pressure from predators and competitors, that is, when ecosystems are not saturated with species (Krassilov 1986; Rasnitsyn 1987) and/or are subjected to an additional influx of energy and nutrients (Vermeij 1995). In addition, the nonmonotone character of evolution leads to the observation that paleontologists rarely record transitional forms. Therefore, comparatively quick changes are followed by long periods of stagnation (even in a changing environment), corresponding to the concept of punctuated equilibria (Gould and Eldredge 1977, 1993).

Second, as an organism becomes more complex, the proportion of prohibited combinations of different features in the space of opportunities also increases (Zavarzin 1969, 1974). Hence, a single evolutionary change in complex organisms is more taxonomically significant, because it overcomes larger gaps of character discontinuity. This means that the difference between related complex organisms is more distinct than between simple ones. Therefore, if we determine the velocity of evolution as the rate of extinction and origination of taxa, we find that evolution accelerates with the growth of organismal complexity (Rasnitsyn 1987). This explains the more rapid evolution observed in higher animals in comparison to primitive ones (Simpson 1944; Schmalhausen 1969).

Consequence 2: Faculty Clearance

Multivariate correspondence offers considerable scope for compensatory interactions. In general, the greater the number of active faculties in organisms, the stronger will be their ability to survive in environments subject to temporal or spatial change (Rasnitsyn 1966, 1971, 1972). The ability of animals to use their organs in variable and, when necessary, novel ways is allowed only because structures are matched loosely to particular behavioral patterns or biologic roles (Gans 1974, 1993); this is the second consequence. In other words, it can be expressed as one of the following statements: Changes of structures and functions are relatively independent; organisms are not rigid systems; biologic construction has a clearance (or lost motion) for every function, and vice versa; functioning conditions imposed by natural selection will permit, within certain limits, selection-neutral changes in structures. Such a "clearance" is depicted in figure A2, B by a loose scroll that determines the correspondence between structure and function. This clearance explains why different methods of achieving the same functional result can be found within taxons of any level (Maslov 1980).

Thus, the principle of adaptive trade-off limits constructional change, whereas the second consequence is just the opposite, a permitting rule (Gans 1974, 1993), and the two principles are mirror images of each other. Taken as a whole, they neither contradict numerous examples of amazing adaptations nor, at the same time, the possession by organisms of many features that are neutral and that can be neither eliminated nor supported by natural selection. Neutral characters can represent the absolute majority of all the taxonomically significant features of living beings (Schmalhausen 1983).

The internal diversity of a taxon is only changed to preserve the archetype of the taxon. The archetype is a historically achieved adaptive trade-off of functions. Moreover, it is evolutionarily stable, because it provides adequacy of organisms to the adaptive zone of the taxon. A change in principal design means a violation of the trade-off claims and subsequent loss of adequacy to the adaptive zone. Therefore, such a change can happen only when a new adaptive zone is available. Occupation of the new zone by organisms results in the formation of a new adaptive trade-off, which is recognized as a new archetype and is interpreted as the appearance of another taxon.

Joint consideration of multivariate correspondence and hierarchic organization leads to the conclusion that the lower the hierarchy, the more abundant are nonadaptive and even antiadaptive structures and functions. By contrast, the higher the hierarchic level of a function, the more adaptive features can be found in structures that fulfill this function. This is the main conclusion of this review.

As a result, faculties that belong to lower levels of morphofunctional hierarchy are to a lesser degree subjects for changing the prohibition set by adaptive trade-off; the worth of their constancy is minimal from the perspective of natural selection. This is symbolized in figure A2B by the increase in faculty clearance toward the lower part of the pyramid. Consequently, one adaptive novelty on the lower levels can be accompanied by acquisition of several nonadaptive features. If they appear to be antiadaptive, other low-level faculties can be changed for compensation. The directions of these changes are mostly determined by the logic of embryonic processes, and an environment is free to select adequate phenotypes from a series of opportunities offered by ontogeny. Even if any particular low-level feature is nonadaptive, a malfunction can be compensated by multiple doubling of particular functions. This consideration fits the early observation that features of major taxa are more often adaptive than those of subordinate groups (Chetverikov 1926; Schmalhausen 1983). Neutral characters never occupy a constant position in the phenotypes of higher taxa. As evolution proceeds, they do not become features specific for major groups.

Optimality and Adequacy

In recent literature, the hypothesis of symmorphosis is extensively discussed: Structural parameters fit functional requirements (cited in Garland and Huey 1987; see also Weibel et al. 1991); structures have the form that satisfies maximum necessities, but does not exceed them (Schmidt-Nielsen 1984; Alexander 1988; Garland and Carter 1994); "no more structure is built and maintained than is required to meet functional demand" (Taylor and Weibel 1981; Taylor et al. 1996). This viewpoing has been supported by data on the bird respiratory system (Liem 1988), liquid transport systems (LaBarbera 1990), and the chain process of oxygen and substrate supply for oxidative muscle metabolism (Weibel et al. 1996) in animals.

The concept of symmorphosis is close, if not equal, to the statement that faculties are optimal (Rosen 1967; Shwartz 1968; Jean 1978; Alexander 1996). Optimality is quite attractive, because it allows calculation and thus an explanation of the structural parameters of organisms; it opens a perspective for predictions, which are still rare enough in biology but constitute the main goal for any normal science.

At first glance, the idea of optimality seems to be supported by the concept of stabilizing selection (Schmalhausen 1949) that minimizes energy waste of the organism. Also, numerous different parameters of plant and animal structures and functions are known to be allometrically related (D'Arcy Thompson 1942; Reeve and Huxley 1945; Colbert 1948; White and Gould 1965; Gould 1966, 1971; McMahon 1973; Kofman 1981a, 1981b, 1986; Dolnik 1982; Swartz and Biewener 1992; Heinrich 1993). Rosen (1967) derived the very fact of allometric relations among parameters directly from the condition of optimality and thus considered allometry as a proof of optimality. However, Rosen's conclusions were later shown to be based on a logical mistake (Kofman 1986). Nevertheless, allometric relations show that systems of organs operate in such a mode that any change of the major parameters leads to a strong decrease of adaptedness, that is, to reduced energy flow through the organism as an element of an ecosystem. The conclusion about optimality of systems is only the next step, since the simplest explanation for stability is that faculties correspond to a local minimum, that is, optimum. So, what in an organism corresponds to a local minimum?

Any allometry within a taxon undermines the presence of a similarity, that is, common traits shared by species of the taxon. Hence, something is kept constant, while organisms within the taxon are speciating. This "something" is an invariant character of the taxon archetype; allometry only reflects constancy of archetype. Archetype-forming general faculties are constant not because they themselves are optimal, but because their optimization would negatively influence other faculties and cause a decrease in adaptedness of the whole organism. This is what we call functional (or external) constraints on evolution, while a new faculty condition that does not affect other functions cannot be achieved owing to developmental (or internal) constraints. Thus, although structures are suboptimal in relation to functions, the phenotype of each species is characterized by the maximum adaptedness that an organism belonging to a particular genus can have within the environment it occupies. In other words, the phenotype is a local extremum of morphospace. This does not mean that the organism is optimal related to the niche it occupies. After all, a taxon can be partly or, what is more rare, completely substituted for in its niche (or adaptive zone) by another taxon of a different general organization, implying that the latter fits the niche better. For instance, mammals active during cool weather and at night replaced reptiles that were then unable to maintain a high body temperature. Also, numerous examples are given at the species level by the history of invasions of animals and plants (Elton 1958) and, at the level of higher taxa, by the very history of life on earth, which shows that low-energy organisms possessing passive functions were gradually substituted by high-energy competitors with more active functions. These competitors appeared to be, in general, more adequate to a vast diversity of niches, except for low-energy environments that did not provide organisms with the amounts of energy sufficient for active functioning (see Vermeij [1987] for an extensive list of examples).

Allometrical correlation, in general, should be interpreted carefully. Data on allometry appear not as lines, but as clouds of dots in the space of investigated parameters. The simplest procedure, and what is normally done, is to calculate the general regression line for the group. However, at the moment it is difficult to say what this line will reflect: the inherent properties of the group under study, a numerical parameter of its archetype, or simply a statistical artifact, whether it is formally significant or not. Typically these regressions are valid only in the range of the data at hand (Schmidt-Nielsen 1984), so the first and the most important problem in the calculation of archetype-describing equations is the selection of the initial data. If all available unsorted figures are lumped for analysis, it is easy to obtain a line lacking any biologic meaning.

Routine statistical approaches are restricted by structuredness of data. Statistics normally demand that the scattering of dots within the cloud is purely random and will a priori consider the causality of all deviations to be insignificant. Still, deviations are exactly what we are looking for whenever the optimality of phenotypes is being analyzed. Whenever the cloud reflects data on the hierarchic system of taxa, it must be structured. Therefore, deviations from the general regression line of species-corresponding dots reflect specificity of natural groups of organisms first and mistakes in measurement last.

Let us return to one of the examples presented under "The Principle of Evolutionary Stabilization of Functions." McMahon (1975) found the lengths and diameters of forelimb bones of Bovidae (antelopes, sheep, goats, and cows) neither smaller nor greater than that demanded by the theory of elastic similarity. Hence, the hypothesis of optimal elasticity gives a quite satisfying description of the size of forelimb bones in Bovidae. In addition, whenever not one but several families of terrestrial mammals are being considered, it becomes evident that the character of interrelationships between parameters of the bones and body mass differs in different taxa (Alexander et al. 1979). This should be surprising, because different groups possess different skeletal constructions. Therefore, the optimum for an element is defined by the construction within which the element works. This circumstance sets strong constraints on our ability to calculate and predict morphologies; having obtained a satisfactory explanation for the shape of elements, we still lack an explanation for the origin of the system as a whole, although this determines the structure of its component elements.

Next, suppose that both sets of data reflect true morphofunctional regularities in mammals. Since the correlation between bone parameters and body mass differs for the Mammalia as a whole class and for subordinate taxa, it is evident that mammalian organization itself sets its own demands on peculiar structural parameters. However, the optima for a family or order coincide with general mammalian optima only at the single point at which the two allometry lines cross.

We can now see that the adaptive radiation undergone by a taxon results at the expense of more or less significant deviations of species' phenotypes from both of the two optima. In reality, it takes place by the deviation from all the optima specific for each higher taxon. Although this is true, it can easily happen that nature contains no single species with parameters that perfectly match the optima defined by archetypal construction of its genus, family, order, and other categories. Ideally that may be the only species in a type or, more directly, the only specimen in nature, as the phenomenon of variability itself prohibits the idea of optimality. Hence, it is more meaningful to speak of adequacy (Gans 1993) than of optimality of archetypes. The concept of adequacy indicates that within a given adaptive zone any particular organism can exist if it can maintain a number of specimens no less than the minimum needed for reproduction. In general, the idea of adequacy corresponds to the statment by Cuvier that "different parts of each being must be co-ordinated in such a way as to render possible the existence of the being as a whole" (Russell 1916).

Although multifunctionality rarely allows good matching of real parameters to precalculated optima, the concept of optimality should not be neglected or forgotten. In general, optima are useful tools for investigation, constituting targets toward which natural selection aims. Optima are extreme manifestations of archetypes, and functional optimality constitutes an important particular case of adequacy. A structure can be strongly stabilized within a systematic group, particularly if its function is critical to the life of most species throughout the group. An example occurs in bovids (McMahon 1975). They may share habitats characterized by extensive ranges and uniform food but also diverse, rapidly running predators. It is even suggested that the savanna biome originated under the strong influence of large herbivores like ungulates and mainly bovids (V. V. Zherikhin, personal communication). Hence bovids have a peculiar kind of highly specialized locomotion. Evidently neither the skeletal nor the muscle arrangement of bovids can be changed, since this would result in a loss of adequacy to their niche. Therefore, speciation in this family (which includes more than a hundred recent species) proceeded at the cost of other functional systems, such as courtship behavior and the accompanying change of horn shape (Young 1981; Colbert and Morales 1990). So, optimality of parameters of limb bones in bovids indicates that these parameters characterize the skeleton at a given body mass; they cannot be considered as traits of species. As a by-product of bone size optimization, bovids lost the ability to operate their extremities in the various manners of other mammals: all the movements are restricted to fore-and-aft direction, and the relative motility of single limb elements is strongly limited (Young 1981). Supposedly, more cases close to the optimum state occur in specialized forms with a reduced amount and diversity of functions.

Studies of optima can sometimes provide important information, because the optima set by different functions are the limits among which the archetype is located. Nevertheless, many optima are too far from what we see in nature. Limits set by such optima evidently cannot guide morphologic research.

Conclusions

General principles of organismal construction permit changes that proceed relatively independently in structures and functions, thereby enabling the manifestation of both purely morphologic and physiologic aspects in organisms. Therefore, necessary and sufficient explanations of organismal features must include selective and morphogenetic considerations, which represent two independent and mutually additive factors of biologic pattern formation.

The pattern formation can be illustrated by a simple analogy of an evolving organism to a traveling car, taking ontogenetic trajectories as roads and natural selection as the driver (fig. A7). Every time environmental conditions change, the driver is free to select one from a number of roads offered by ontogeny. The paths chosen are adaptive; those unchosen are not. The latter

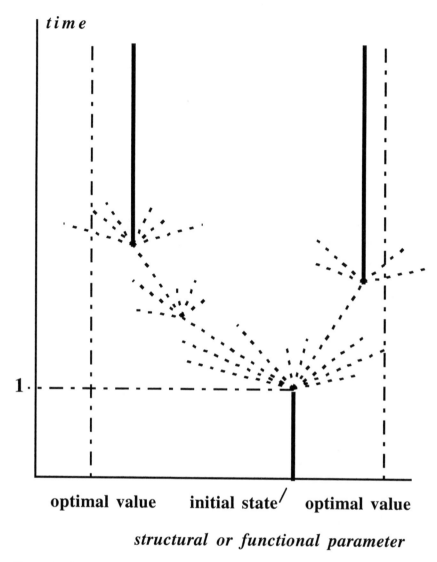

time

1

optimal value   initial state/   optimal value

structural or functional parameter

Figure A7. General view of evolution. 1, the moment of environmental change.

are always more numerous; it is a well-known paleontologic fact that the diversity of a taxon abruptly increases before its extinction (Chernykh 1986; Rasnitsyn 1987).

Of course, the traveling car analogy is very simple, since our roads are stable in time and can be depicted on a map; it may seem that elaboration of an evolutionary prognosis only requires learning of the "map." In fact, the spectrum of possible forms determined by morphogenesis is generated partly de novo during each elementary evolutionary change; thus it cannot be predescribed in the long run.

The complementarity of the two factors of pattern formation engenders a strong uncertainty of morphologic explanations. This uncertainty is not absolutely insuperable, since the factors impose different effects on form, depending on the level of morphofunctional hierarchy: high level (systemic) features are explained primarily as adaptations, whereas low-level characters (elementary) are instead interpreted morphogenetically.

Regrettably, I am not equipped to provide full empirical support to this entire set of statements. I tried to corroborate it in this book, using examples from insect wings and flight. I argued that a reliable selectionist explanation can be given to the features that characterize the general construction of the flight system (the number of operating wings, their relative size and type of interactions while flying). These features characterize ordinal taxa. By contrast, the morphologic peculiarities found on the wings of different small groups mostly lack good selectionist interpretation. These materials agree with the general evolutionary considerations of this appendix, namely that adaptive properties of morphologic features decrease from the top to the bottom of the organismal constructional hierarchy.

However, this definitely does not imply that adaptive features are absent at the lower levels of morphofunctional hierarchy and that ontogenetically caused traits cannot be found on the top level. They doubtless can. For example, in the venation of insect wings, a strengthened radius and cubitus anterior—the main levels controlling wing movement—are clearly adaptive features. Hence the general conclusion should not be treated as an absolute rule. I only want to emphasize the relatively low contribution of adaptive and nonadaptive characters in the total complex of peculiarities of corresponding hierarchic levels. Thus, a new feature that is encountered while studying the morphology of an organism should first be considered from the viewpoint of a level-corresponding hypothesis, that is, adaptive for a system of organs and morphogenetic for an element of an organ. Only after the starting hypothesis is falsified should an alternative explanation be examined; this is simply a time-saving approach.

Two apparent contradictions to the proposed ideas must be discussed. The first is: "One can speak of minimal changes in the upper levels of organization only if he knows how to distinguish the upper and lower levels. No approach to functional hierarchy is seen besides this one: the more taxonomically and evolutionarily stable is the function, the more important it is for the organism, thus the higher its rank. As a result, the rank of a function is determined based on the principle of minimality, and if you then try to corroborate it with the help of the same rank system, a never-ending circle is obtained" (A. P. Rasnitsyn, personal communication).

Indeed, differentiation between levels of morphofunctional hierarchy is much easier in theory than in reality. It is an eternal philosophical question: what is whole and what is part? I daresay that, concerning this problem, characters used to describe a whole determine interrelationships between its parts and correspond to several organs simultaneously. Accordingly, it is not surprising that only these characters appear to be stable under comparison of different groups, common to the majority of the taxon members and in sum give Owen's archetype of the taxon. The same is suggested by experience of taxonomy and phylogenetics, in which greater weight has traditionally been given to complex features (Mayr 1969), such as the gross structure of the main organs or the type of symmetry (Rasnitsyn 1996).

The features of construction that determine the coordination of organs are functional by definition, and functions at the organismal level (e.g., the ability to feed, move, breed) are always adaptive. Hence, the higher the taxon level, the more adaptive the features characterizing the taxon: insect orders are classified according to functional systems—types of mouthparts and flight apparatus—whereas the animal and plant realms differ only by the source of consumed energy, a purely physiologic character. So coincidence of megataxa with living forms seems quite natural and is not a disadvantage of the existing megaclassification, as is sometimes thought (see Koussakin and Drozdov 1994).

Nevertheless, functional and morphologic hierarchies do not completely coincide with taxonomic systems: not all the features characterizing higher taxa obligatorily occupy high hierarchic levels, because features exist that are directly connected with major archetype parameters and thus are quite stable for taxonomic use. For example, the heat-insulating outer coverings of birds, mammals, and lepidopterans are associated with the ability of these organisms to maintain constant body temperature. The very capacity for thermoregulation could not evolve without association with the appearance of heat-insulating outer structures. The presence of the covering is reliably associated with taxonomy, although from a morphologic viewpoint it constitutes simply a derivative of epidermis. In addition, some of the major insect orders (Papilionida, Phryganeida, Vespida) are reliably characterized by the anal loop, which is very taxonomically stable and originates owing to wing narrowing, but plays no adaptive role.

The second strong contradiction is given by the consideration that in agreement with the principle of minimal change, the structures of the lower levels of the morphofunctional hierarchy must be more diverse, since they undergo more complete reconstruction during each step of evolution. On the one hand, this is supported by tissue incompatibility observed even in close species. On the other, the general organization of animal tissues is much more uniform than that of the fauna (Zavarzin 1934, 1950, 1953): the number of tissue types can be compared with the number of major but not medium and minor taxa.

Ways to remove this contradiction are unclear at the moment. It can only be stated that, from the viewpoint of the presented approach, the interpretation of Zavarzin (1950, 1953) cannot be adopted, because it involves the "unity of form and function" understood as the necessity of functional traits to fit mechanical conditions, thus implying the origin of the same form under each given function. At the same time, the stability of tissue structure appears relatively normal from the perspective of comparative embryology, because in ontogeny, according to von Baer's rule, tissues are differentiated earlier than morphologic structures and therefore must be more similar.

Note also, however, that hierarchy constitutes an extremely simple concept in respect to the problems of describing biologic organization. A more complex scheme is given by heterarchy, a "constellation" of relatively independent and interrelated functional parts (Goodwin 1976). Heterarchy lacks a predominant part, since any of the functional systems can become limiting under particular conditions (Rasnitsyn 1987).

The heterarchic scheme is obtained if we overlap the multivariate correspondence and a hierarchic pyramid without a top. The upper surface of this pyramid corresponds to the heterarchic system, and hence hierarchy is an adequate approach to describing biologic complexity, although it is insufficient when taken alone, since the multivariate correspondence between structures and functions is also necessary. In principle, the top of the morphologic hierarchy is the organism itself, whereas the top of the physiologic pyramid is the main function of the organism, that is, *life.*

References

Alberch P. 1980. Ontogenesis and morphological diversification. Am. Zool. 20:653–67.

———. 1981. Convergence and parallelism in foot morphology in the Neotropical salamander genus *Bolithoglossa*. 1. Function. Evolution 35(1):84–100.

Alberch P., Gale E. A. 1985. A developmental analysis of an evolutionary trend: digital reduction in amphibians. Evolution 39(1):8–23.

Aldridge H. D. J. N. 1986. Kinematics and aerodynamics of the greater horseshoe bat, *Rhinolophus ferrumequinum,* in horizontal flight at various flight speed. J. Exp. Biol. 126:479–97.

Alexander D. E. 1984. Unusual phase relationships between the forewings and hindwings in flying dragonflies. J. Exp. Biol. 109:379–83.

———. 1986. Wind tunnel studies of turns by flying dragonflies. J. Exp. Biol. 122:81–98.

Alexander R. D., Brown W. L. 1963. Mating Behavior and the Origin of Insect Wings. Ann Arbor, Mich. 19 p.

Alexander R. McN. 1968. Animal Mechanics. Seattle, Wash. 339 p.

———. 1986. Three kinds of flying in animals. Nature 321(6066):113–14.

———. 1988. The scope and aims of functional and ecological morphology. Neth. J. Zool. 38(1):3–22.

———. 1996. Optima for Animals. N.J.: Princeton Univ. Press. 169 p.

Alexander R. McN., Jayes A. S., Maloiy G. M. O., Wathuta E. M. 1979. Allometry of the limb bones of mammals from shrews (*Sorex*) to elephant (*Loxodonta*). J. Zool. 189(3):305–14.

Antonova O. A., Brodsky A. K., Ivanov V. D. 1981. Wingbeat kinematics in five insect species. Zool. Zhurn. 60(4):506–18 (in Russian).

Arbas E. A. 1986. Control of hindlimb posture by wind-sensitive hairs and antennae during locust flight. J. Comp. Physiol. 159(6):849–57.

Arnett R. H. 1993. American Insects. A Handbook of Insects of America North of Mexico. Gainesville, Fla: Sandhill Crane Press. 850 p.

Arnold S. J., Alberch P., Csanyi V., Dawkins R. C., Emerson S. B., Fritzsch B., Horder T. J., Maynard Smith J., Starck M. J., Vrba E. S., Wagner G. P., Wake D. B. 1989. How do complex organisms evolve? In: Wake D. B., Roth G., eds. Complex Organismal Functions: Integration and Evolution in Vertebrates. New York: Wiley. pp. 403–33.

Arshavsky I. A. 1968. The principle of function change in the ontogeny of mammals. Zhurn. Obstchej Biologii 29(1):78–87 (in Russian).

Ashley H. 1974. Engineering Analysis of Flight Vehicles. New York: Dover. 386 p.

Ashworth J., Luttges M. 1986. Comparisons in three-dimensionality in the unsteady flows elicited by straight and swept wings. American Institute of Aeronautics and Astronautics, paper 86-2280. 10 p.

Ashworth J., Mouch T., Luttges M. 1988. Application of forced unsteady aerodynamics to a forward swept wing X-29 model. American Institute of Aeronautics and Astronautics, paper 88-0563. 12 p.

Averof M., Cohen S. M. 1997. Evolutionary origin of insect wings from ancestral gills. Nature 385(6617):627–30.

Azuma A. 1992. The Biokinetics of Flying and Swimming. Tokyo: Springer. 266 p.

Azuma A., Azuma S., Watanabe T., Furuta T. 1985. Flight mechanics of a dragonfly. J. Exp. Biol. 116:79–108.

Azuma A., Watanabe T. 1988. Flight performance of a dragonfly. J. Exp. Biol. 137:221–52.

Babcock J. M. 1985. An Alaskan record for mountain midges (Diptera: Deuterophlebiidae) with notes on larval habitat. Entomol. News 96(5):209–10.

Baker J. M. 1965. Flight behaviour in some anobiid beetles. In: Proc. 12th Int. Congr. Entomol.; 8–16 July 1964; London. pp. 319–20.

Baker P. S. 1979. The wing movements of flying locusts during steering behaviour. J. Comp. Physiol. A 131(1):49–58.

Baker P. S., Cooter R. J. 1979a. The natural flight of the migratory locust, Locusta migratoria L. 1. Wing movements. J. Comp. Physiol. A 131(1):79–87.

————. 1979b. The natural flight of the migratory locust, Locusta migratoria L. 2. Gliding. J. Comp. Physiol. A 131(1):89–94.

Baker P. S., Cooter R. J., Chang P. M., Hashim H. B. 1980. The flight capabilities of laboratory and tropical field populations of the brown planthopper, Nilaparvata lugens (Stal) (Hemiptera: Delphacidae). Bull. Entomol. Res. 70(4):589–600.

Baker P. S., Gewecke M., Cooter R. J. 1981. The natural flight of the migratory locust, Locusta migratoria L. III. Wing-beat frequency, flight speed and attitude. J. Comp. Physiol. A 141:233–37.

Banerjee S. 1988. Organization of the cuticle in Locusta migratoria Linnaeus, Tropidacris cristata Linnaeus and Romalia microptera Beauvais (Orthoptera, Acrididae). Int. J. Insect Morphol. Embryol. 17:313–26.

Barber S. B., Pringle J. W. S. 1965. The functional organization of the flight system in belostomatid bugs (Heteroptera). In: Proc. 12th Int. Congr. Entomol.; 8–16 July 1964; London. pp. 185–86.

Bartholomew G. A., Casey T. M. 1978. Oxygen consumption of moths during rest, preflight warm-up, and flight in relation to body size and wing morphology. J. Exp. Biol. 76:11–25.

Bartholomew G. A., Vleck D., Vleck C. M. 1981. Instantaneous measurements of oxygen consumption during pre-flight warm-up and post-flight cooling in sphingid and saturniid moths. J. Exp. Biol. 90:17–32.

Batchelor G. K. 1967. An Introduction to Fluid Dynamics. Cambridge: Cambridge Univ. Press. 615 p.

Bechert D. W., Hoppe G., Reif W.-E. 1985. On the drag reduction of the shark skin. American Institute of Aeronautics and Astronautics, paper 546. 18 p.

Becker E. G. 1952. On the problem of insect wing origin. Part 1. Precursors of insect wing. Vestnik MGU, no. 9:59–68 (in Russian).

———. 1966. On the origin and development of the insect wing. In: Teoria morphologicheskoj evoliutsii nasekomykh. Moscow: Izd-vo MGU. pp. 174–224 (in Russian).

Beklemishev V. N. 1994. Metodologia Sistematiki [The Methodology of Taxonomy]. Moscow: KMK Scientific. 252 p. (in Russian).

Belintsev B. N. 1991. Fizicheskie Osnovy Biologicheskogo Formoobrazovania [Physical Basis of Biological Pattern Formation]. Moscow: Nauka. 256 p. (in Russian).

Belotserkovsky S. M., Gulyaev V. V., Nisht M. I. 1974a. Towards the study of insect and bird flight. Doklady Academii Nauk SSSR 219:567–70 (in Russian).

———. 1974b. On the mechanism of generation of normal force by flap of wings. In: Izbrannye Problemy Prikladnoj Mechaniki. Moscow: Nauka. pp. 97–102 (in Russian).

Belotserkovsky S. M. and Nisht M. I. 1978. Otryvnoe i bezotryvnoe obtekanie tonkikh kryliev idealnoj zhidkostju [Stalling and Non-stalling Ideal Flow around Thin Aerofoil]. Moscow: Nauka Publishing House (in Russian).

Beloussov L. V. 1975. Parametric system of Thecaphora hydroids and possible ways of genetical regulation of their inter-species differences. Zhurn. Obstchej Biologii 36(5):654–62 (in Russian).

———. 1980. Vvedenije v Obstchuju Embriologiju [Introduction to General Embryology]. Moscow: Izd-vo MGU. 212 p. (in Russian).

———. 1987. Biologicheskij Morfogenez [Biological Morphogenesis]. Moscow: Izd-vo MGU. 238 p. (in Russian).

Bennet-Clark H. C. 1986. The flight of the dipteran fly. Nature 321(6069):468–69.

Bennett L. 1976. Induced airflow created by large hovering beetles. Ann. Entomol. Soc. Am. 69(6):985–90.

Berg L. S. 1977. Trudy po Teorii Evoliutsii [Papers on the Theory of Evolution]. Leningrad: Nauka. 388 p. (in Russian).

Berg R. L. 1993. Genetika i Evoliutsia [Genetics and Evolution]. Novosibirsk: Nauka. 284 p. (in Russian).

Betts C. R. 1986a. The comparative morphology of the wings and axillae of selected Heteroptera. J. Zool. B 1(2):255–82.

———. 1986b. Functioning of the wings and axillary sclerites of Heteroptera during flight. J. Zool. B 1(2):283–301.

———. 1986c. The kinematics of Heteroptera in free flight. J. Zool. B 1(2):303–15.

Betts C. R., Wootton R. J. 1988. Wing shape and flight behaviour in butterflies (Lepidoptera: Papilionoidea and Hesperioidea): a preliminary analysis. J. Exp. Biol. 138:271–88.

Biewener A. A. 1982. Bone strength in small mammals and bipedal birds: do safety factors change with body size? J. Exp. Biol. 98:289–301.

————. 1983. Locomotory stresses in the limb bones of two small mammals: the ground squirrel and chipmunk. J. Exp. Biol. 103:131–54.

Bigelow R. S. 1958. Classification and phylogeny. Syst. Zool. 7:49–59.

Billberg G. J. 1820. Enumeratio Insectorum in Museo. Gadelianis. 138 p.

Birket-Smith S. J. R. 1984. Prolegs, legs and wings of insects. Entomograph. 5:1–128.

Bliakher L. Ya. 1976. Problemy Morfologii Zhivotnykh [The Problems of Animal Morphology]. Moscow: Nauka. 360 p. (in Russian).

Blondeau J. 1981. Aerodynamic capabilities of flies, as revealed by a new technique. J. Exp. Biol. 92:155–64.

Bocharova-Messner O. M. 1968. The principles of pterothorax ontogeny in Polyneoptera as related to the problem of origin and evolution of insect flying systems. In: Voprosy Funktsional'noj Morfologii i Embriologii Nasekomykh [Problems of Insect Functional Morphology and Embryology]. Moscow: Nauka. pp. 3–26 (in Russian).

————. 1979a. Surface aerodynamic peculiarities of insect wings. In: Sokolov V. E., ed. Adaptivnye Svojstva Epitelija i ego Proizvodnykh [Adaptive Properties of Epithelium and Its Derivatives]. Moscow: Nauka Publishing House. pp. 69–106 (in Russian).

————. 1979b. Insect wings as flight organs. In: Doklady na 30om Ezhegod. Chtenii Pamiati N.A.Kholodkovskogo. Leningrad: Nauka. pp. 3–40 (in Russian).

————. 1982. Morphological model of life-supporting system as revealed from the example of insect flapping plane. In: Vorobjova E. I., ed. Problemy Razvitija Morfologii Zhivotnykh [Problems of Development of Animal Morphology]. Moscow: Nauka. pp. 128–39 (in Russian).

Bocharova-Messner O. M., Aksiuk T. S. 1981. In-flight formation of tunnel by butterfly (Lepidoptera, Rhopalocera) wings. Dokl. Akad. Nauk SSSR 260(6):1490–93 (in Russian).

Bocharova-Messner O. M., Dmitriev A. E. 1984. Morpho-functional analysis of dragonfly wing venation (as revealed from the data of high-speed cine filming and scanning microscopy). In: 9 Sjezd Vsesojuznogo Entomol. O-va. Kiev, Okt. 1984. Tez. dokl. Vol. 1. p. 65 (in Russian).

Bock W. J., von Wahlert G. 1965. Adaptation and the form-function complex. Evolution 19(3):269–99.

Boettiger E. G., Furshpan E. 1952. The mechanics of flight movements in Diptera. Biol. Bull. Woods Hole 102:200–211.

Bohr N. 1971. Selected scientific papers. Vol. 2 Moscow: Nauka. (in Russian).

Bolton H. 1925. Insects from the coal measures of Commentry. Brit. Mus. (Nat. Hist.) Fossil Insects 2:1–56.

Borin A. A. 1987. On take-off capacities of flying animals. Dokl. Akad. Nauk SSSR 293(5):1256–58 (in Russian).

Borin A. A., Kokshaysky N. V. 1982. On the problem of the operational mode of the bird wings. Dokl. Akad. Nauk SSSR 266(3):726–29 (in Russian).

Börner C. 1904. Zur Systematik der Hexapoden. Zool. Anz. 27:511–33.

Borror D. J., Triplehorn C. A., Johnson N. F. 1992. Study of Insects. Fort Worth: Harcourt Brace Coll. Publishers. 875 p.

Boudreaux H. B. 1979. Arthropod phylogeny with special references to insects. New York: Wiley. 320 p.

Brackenbury J. 1990. Wing movements in the bush-cricket *Tettigonia viridissima* and the mantis *Ameles spallanziana* during natural leaping. J. Zool. 220:593–602.

———. 1991a. Wing kinematics during natural leaping in the mantids *Mantis religiosa* and *Iris oratoria*. J. Zool. 223:341–56.

———. 1991b. Kinematics of take-off and climbing flight in butterflies. J. Zool. 224:251–70.

———. 1992. Insects in Flight. London: Blandford. 192 p.

Brackenbury J. H. 1994. Hymenopteran wing kinematics: a qualitative study. J. Zool. 233:523–40.

Brauer F. 1885. Systematische-zoologische Studien. Sitzingsberichte der Kaiserisch Akademie der Wissenschaften. Vol. 91; pp. 237–417.

Brodsky A. K. 1971. Experimental studies of flight of mayfly *Ephemera vulgata* L. (Ephemeroptera). Entomol. Obozr. 50(1):43–50 (in Russian, published in English in Entomol. Rev. [Wash.] 1971; 50(1):25–28).

———. 1974. Evolution of flight apparatus of mayflies (Ephemeroptera). Entomol. Obozr. 53(2):291–302 (in Russian, published in English in Entomol. Rev. [Wash.] 1974; 53(2):35–43).

———. 1975. Wingbeat kinematics of mayflies and the analysis of the mechanism of flight power regulation. Zool. Zhurn. 54(2):209–20 (in Russian).

———. 1979a. The evolution of wing apparatus in stone flies (Plecoptera). Part 2. Functional morphology of the wing hinge, cuticle skeleton and musculature. Entomol. Obozr. 58(4):705–15 (in Russian, published in English in Entomol. Rev. [Wash.] 1979; 58(4):16–26).

———. 1979b. The origin and early stages of evolution of the insect wing apparatus. In: Doklady na 30om ezhegod. chtenii pamiati N. A. Kholodkovskogo. Leningrad: Nauka. pp. 41–78 (in Russian).

———. 1981a. Aerodynamical peculiarities of insect flight. 4. Data on mayfly *Ephemera vulgata* flight. Vestnik LGU, no. 15:12–18 (in Russian).

———. 1981b. The evolution of wing apparatus of stone flies (Plecoptera). Part 3. In-flight wing deformation in stone fly *Isogenus nubecula* Newman. Entomol. Obozr. 60(3):523–34 (in Russian, published in English in Entomol. Rev. [Wash.] 1981; 60(3): 25–36).

———. 1982. Evolution of wing apparatus in stone flies (Plecoptera). Part 4. Wingbeat kinematics and general conclusions. Entomol. Obozr. 61(3):491–500 (in Russian, published in English in Entomol. Rev. [Wash.] 1982; 61(3):34–43).

———. 1984. Aerodynamic peculiarities of insect flight. 5. Vortex formation during flapping flight. Vestnik LGU, no. 21. Ser. Biol., issue 4. pp. 23–28 (in Russian).

———. 1985. Insect wings movement during horizontal steady flight (a comparative approach). Entomol. Obozr. 64(1):33–50 (in Russian, published in English in Entomol. Rev. [Wash.] 1985; 64(3):56–73).

———. 1986a. The flight of giant stone fly *Allonarcys sachalina* (Plecoptera, Pteronarcyidae) and the analysis of the mechanism of insect wings supination. Zool. Zhurn. 65(3):349–60 (in Russian).

———. 1986b. Insect flight at high wingbeat frequency. Entomol. Obozr. 65(2):269–79 (in Russian).

———. 1987. Structure and functional significance of veins and furrows on insect wings. In: Morfologicheskije Osnovy Filogenii Nasekomykh [Morphological Bases of Insect Phylogeny]. Leningrad: Nauka. pp. 4–19 (in Russian).

———. 1988. Mekhanika poliota nasekomykh i evoliutsia ikh krylovogo apparata [Mechanics of Flight and Evolution of Flying System in Insects]. Leningrad: University Press. 208 p. (in Russian).

———. 1990. Experimental studies of the flight of the peacock butterfly *Inachis io* L. (Lepidoptera, Nymphalidae). Zool. Zhurn. 69(9):39–50 (in Russian).

———. 1991. Vortex formation in the tethered flight of the peacock butterfly and some aspects of insect flight evolution. J. Exp. Biol. 161:77–92.

———. 1994. The Evolution of Insect Flight. New York: Oxford Univ. Press. 229 p.

Brodsky A. K., Grodnitsky D. L. 1985. The aerodynamics of tethered flight of the european skipper *Thymelicus lineola* Ochs. (Lepidoptera, Hesperiidae). Entomol. Obozr. 64(3):484–92 (in Russian, published in English in Entomol. Rev. [Wash.] 1986; 65(3):60–69).

Brodsky A. K., Ivanov V. D. 1983a. The visualization of air flow around flying insects. Dokl. AN SSSR 271(3):742–45 (in Russian).

———. 1983b. The functional assessment of the construction of insect wings. Entomol. Obozr. 62(1):48–64 (in Russian, published in English in Entomol. Review [Wash.] 1983; 62(1):35–52).

———. 1984. The role of vortices in insect flight. Zool. Zhurn. 63(2):197–208 (in Russian).

Brodsky A. K., Ivanov V. P. 1974. The aerodynamical peculiarities of insect flight. 2. Flow spectra. Vestnik LGU, no. 3:16–21 (in Russian).

———. 1975. The aerodynamical peculiarities of insect flight. 3. Flow around the wings of the mayfly *Ephemera vulgata* L. (Ephemeroptera). Vestn. LGU, no. 3:7–10 (in Russian).

Brodsky A. K., Vorobjov N. N. 1990. Gliding of butterflies and the role of the wing scale covering in their flight. Entomol. Obozr. 69(2):241–56 (in Russian).

Brodsky K. A. 1930. Zur Kenntnis der Wirbellösenfauna der Bergströme Mittelasiens. 2. *Deuterophlebia mirabilis* Edw. Ztschr. Morphol. Okol. Tiere 18(1/2):289–321.

Brongniart C. 1893. Recherches pour servir à l'histoire des insectes fossiles des temps primaires, précédées d'une étude sur la nervation des ailes des insectes. Thèse présentée à la Faculté des Sciences de Paris. No. 821. 495 p.

Brues C. T., Melander A. L. 1932. Key to the families of North American insects. An introduction to the classification of insects. Privately printed. Boston & Pullman. 140 p.

Brullé A. 1832. Classe insectes: expédition scientifique de la Morée. Vol. 3, no. 1. Paris: Levrault. pp. 64–395.

Brunner K. 1882. Prodromus der europäischen Orthoptèren. Leipzig: Engelmann. 466 p.

Buckholz R. H. 1981. Measurements of unsteady periodic forces generated by the blowfly flying in a wind tunnel. J. Exp. Biol. 90:163–73.

Burmeister H. G. R. 1835. Handbuch der Entomologie. Table 2, fig. 1. Berlin: Reimer. 400 p.

Butler P. J., Woakes A. J. 1980. Heart rate, respiratory frequency and wing beat frequency of free flying barnacle geese, *Branta leucopsis*. J. Exp. Biol. 85:213–26.

Bykhovskii B. E., Vedenov M. F., Kremiansky V. I., Shatalov A. T., eds. 1972. Razvitije Kontseptsii Strukturnykh Urovnej v Biologii [Development of the Concept of Structural Levels in Biology]. Moscow: Nauka. 392 p. (in Russian).

Byrne D. N., Buchmann S. L., Spangler H. G. 1988. Relationship between wing loading, wingbeat frequency and body mass in homopterous insects. J. Exp. Biol. 135:9–23.

Camhi J. M. 1970. Yaw-correcting postural changes in locusts. J. Exp. Biol. 52(3):519–31.

Caple G., Balda R. P., Williams W. R. 1983. The physics of leaping animals and the evolution of preflight. Am. Natural 121(4):455–76.

Carle F. L. 1982. Thoughts on the origin of insect flight. Entomol. News 93(5):159–72.

Carpenter F. M. 1992. Treatise on Invertebrate Paleontology. Pt. R. Arthropoda 4. Vol. 3. Superclass Hexapoda. Geological Society of America, Boulder, Colo., and Univ. of Kansas, Lawrence, Kans. 655 pp.

Carpenter R. E. 1985. Flight physiology of flying foxes, *Pteropus poliocephalus*. J. Exp. Biol. 114:619–47.

Casey T. M. 1981. Insect flight energetics. In: Herreid C. F., Fourtner C. R., eds. Locomotion and Energetics in Arthropods. New York: Plenum. pp. 419–52.

———. 1989. Oxygen consumption during flight. In: Goldsworthy G. J., Wheeler C. H., eds. Insect Flight. Boca Raton: CRC Press. pp. 257–72.

Casey T. M., May M. L. 1983. Morphometrics, wing stroke frequency and energy metabolism of euglossine bees during hovering flight. In: Nachtigall W., ed. BIONA Rept. Vol. 1. pp. 1–10.

Casey T. M., May M. L., Morgan K. R. 1985. Flight energetics of euglossine bees in relation to morphology and wing stroke frequency. J. Exp. Biol. 116:271–90.

Chadwick L. E. 1940. The wing motion of the dragonfly. Brooklyn Entomol. Soc. Bull. 35:109–12.

Chai P., Chen J. S. C., Dudley R. 1997. Transient hovering performance of hummingbirds under conditions of maximal loading. J. Exp. Biol. 200:921–29.

Chai P., Dudley R. 1995. Limits to vertebrate locomotor energetics suggested by hummingbirds hovering in heliox. Nature 377:722–25.

———. 1996. Limits to flight energetics of hummingbirds hovering in hypodense and hypoxic gas mixtures. J. Exp. Biol. 199:2285–95.

Chai P., Harrykissoon R., Dudley R. 1996. Hummingbird hovering performance in hyperoxic heliox: effects of body mass and sex. J. Exp. Biol. 199:2745–55.

Chaikovsky Yu. V. 1990. Elementy Evoliutsionnoj Diatropiki. [Elements of Evolutionary Diatropics]. Moscow: Nauka. 272 p. (in Russian).

————. 1994. Transformation of diversity. Khimija i Zhizn. No. 1, pp. 20–29 (in Russian).

Chance M. A. C. 1975. Air flow and the flight of the noctuid moth. In: Swimming and Flight in Nature. Vol. 2. New York: Plenum. pp. 829–43.

Chebanov S. V. 1977. The theory of classification and the methodology of classifying. In: Nauchno-Tekhnicheskaja Informatsia. Ser. 2. Informatsionnye Protsessy i Systemy. Moscow: VINITI. No. 10, pp. 1–10 (in Russian).

————. 1984. Insight into form in natural history and the baselines of general morphology. In: Organilise Voormi Teooria [Theory of Organic Form]. Tartu: Izd-vo AN ESSR. pp. 25–41 (in Russian).

Chernykh V. V. 1986. Problema Tselostnosti Vysshykh Taksonov [The Problem of Integrity of Higher Taxa]. Moscow: Nauka. 144 p. (in Russian).

Chetverikov S. S. 1926. On some points of the evolutionary process from the viewpoint of modern genetics. Zhurn. Experim. Biologii Ser. A 2(1):3–54 (in Russian).

Chikatunov V. I., Denisova A. A. 1988. On the spatial structure and dynamics of phenofunds of alpine populations of rape leaf beetle in alpine Tajikistan. In: Ekologia Populiatsij [Population Ecology]. Tez. Dokl. Vsesojuz. Symp.; 4–6 Oct. 1988; Novosibirsk. Moscow. Pt. 1 (in Russian).

Chikatunov V. I., Kriukova V. I. 1985. Comparative phenetic analysis of wing venation of two species of long-horned beetles *Chlorophorus faldermanni* Fald. and *Xylotrechus namanganensis* Heyd. (Coleoptera, Cerambycidae) from Middle Asia. In: Fenetika Populiatsij [Population Phenetics]. Moscow: pp. 146–47 (in Russian).

Choe J. C. 1992. Zoraptera of Panama with a review of the morphology, systematics, and biology of the order. In: Quintero D., Aiello A., eds. Insects of Panama and Mesoamerica. Selected Studies. Oxford: Oxford Univ. Press. pp. 249–56.

Church N. S. 1960. Heat loss and the body temperatures of flying insects. 2. Heat conduction between the body and its loss by radiation and convection. J. Exp. Biol. 37(1):186–212.

Clairville J., de 1798. Helvetische Entomologie: oder, Verzeichniss der schweizerischen Insekten nach einer neuen Methode geordnet: mit Beschreibungen und Allildungen. Table 1. Zurich: Orell. 149 p.

Clench H. K. 1966. Behavioral thermoregulation in butterflies. Ecology 47(6):1021–34.

Cloupeau M., Devillers J. F., Devezeaux D. 1979. Direct measurements of instantaneous lift in desert locust, comparison with Jensen's experiments on detached wings. J. Exp. Biol. 80:1–15.

Cockbain A. J. 1961. Water relationships of *Aphis fabae* Scop. during tethered flight. J. Exp. Biol. 38:175–80.

Colbert E. H. 1948. Evolution of the horned dinosaurs. Evolution 2:143–63.

Colbert E. H., Morales M. 1990. Evolution of the Vertebrates. A History of the Backboned Animals through Time. 4th ed. New York: Wiley. 470 p.

Collett T. S., Land M. F. 1975. Visual control of flight behaviour in the hoverfly, *Syritta pipiens* L. J. Comp. Physiol. 99:1–66.

Comstock J. H. 1918. The Wings of Insects. New York: Ithaca. pp. 1–430.

Comstock J. H., Comstock A. B. 1895. A Manual for the Study of Insects. New York: Comstock Pub. Co. Ithaca. 701 p.

Comstock J. H., Needham J. G. 1898. The wings of insects. Am. Natural 32:43–903.

———. 1899. The wings of insects. Am. Natural 33:118–858.

Cooter R. J. 1973. Flight and landing posture in hoppers of Schistocerca gregaria (Forsk.). Acrida. 2:307–17.

———. (1979). Visually induced yaw movements in the flying locust, Schistocerca gregaria (Forsk.). J. Comp. Physiol. 131(1):67–78.

Cooter R. J., Baker P. S. 1977. Weis-Fogh clap and fling mechanism in Locusta. Nature 269(5623):53–54.

Crampton G. C. 1915. The thoracic sclerites and the systematic position of Grylloblatta campodeiformes Walker, a remarkable annected orthopteroid insect. Entomol. News and Proc. Entomol. Section Acad. Sci. Philadelphia. Vol. 26. pp. 337–51.

Cullen M. J. 1974. The distribution of asynchronous muscle in insects with particular reference to the Hemiptera: an electron microscope study. J. Entomol. London A 49:17–41.

Cullis N. A., Hargrove J. W. 1972. An automatic device for the study of tethered flight in insects. Bull. Entomol. Res. 61(3):533–37.

Dalton S. 1975. Borne on the Wind. London: Chatto & Windus. 160 p.

Danforth B. N. 1989. The evolution of hymenopteran wings: the importance of size. J. Zool. 218(2):247–76.

Danforth B. N., Michener C. D. 1988. Wing folding in the Hymenoptera. Ann. Entomol. Soc. Am. 81:342–50.

D'Arcy Thompson W. 1942. Growth and Form. Cambridge: Cambridge Univ. Press. 1116 p.

Darwin C. 1872. The Origin of Species by Means of Natural Selection. 6th ed. London: J. Murray.

David C. T. 1978. The relationship between body angle and flight speed in free-flying Drosophila. Physiol. Entomol. 3(3):191–95.

Davis D. R. 1978. A revision of the North American moths of the superfamily Eriocranioidea with the proposal of a new family, Acanthopteroctetidae (Lepidoptera). Smithsonian Contrib. Zool. 251:1–131.

Davis D. R., Faeth S. H. 1986. A new oak-mining eriocraniid moth from southeastern United States (Lepidoptera: Eriocraniidae). Proc. Entomol. Soc. Washington 88(1):145–53.

DeGeer C. 1773. Mémoirs pour servir l'histoire des insectes. Table 3. Stockholm. 696 p.

Demoll R. 1918. Der Flug der Insekten und der Vögel. Jena.

Dickinson M. H. 1994. The effects of wing rotation on unsteady aerodynamic performance at low Reynolds numbers. J. Exp. Biol. 192:179–206.

———. 1996. Unsteady mechanisms of force generation in aquatic and aerial locomotion. Am. Zool. 36(6):537–54.

Dickinson M. H., Götz K. G. 1993. Unsteady aerodynamic performance of model wings at low Reynolds numbers. J. Exp. Biol. 174:45–64.

————. 1996. The wake dynamics and flight forces of the fruit fly *Drosophila melanogaster*. J. Exp. Biol. 199:2085–2104.

Dickinson M. H., Lehmann F. O., Götz K. G. 1993. The active control of wing rotation by *Drosophila*. J. Exp. Biol. 182:173–89.

Dogel' V. A. 1954. Oligomerizatsija Gomologichnykh Organov kak Odin iz Glavnykh Putej Evoliutsii Zhivotnykh [Oligomerization of Homological Organs as One of the Main Trends in Animal Evolution]. Moscow: Izd-vo AN SSSR 368 p.

————. 1981. Zoologija Bespozvonochnykh [Invertebrate Zoology]. Moscow: Visshaja Shkola. 606 p. (in Russian).

Dohrn A. 1875. Der Ursprung der Wirbelthiere und das Princip des Funktionswechsels. Leipzig.

Dolnik V. R. 1982. The allometry of morphology, functions and energetics of homoiothermic animals and its physiological control. Zhurn. Obstchej Biologii 43(4):435–54 (in Russian).

Dorsett D. A. 1962. Preparation for flight by hawk-moths. J. Exp. Biol. 39(4):579–88.

Douglas M. M. 1981. Thermoregulatory significance of thoracic lobes in the evolution of insect wings. Science 211(4477):84–86.

Dudley R. 1990. Biomechanics of flight in neotropical butterflies: morphometrics and kinematics. J. Exp. Biol. 150:37–53.

————. 1991a. Biomechanics of flight in neotropical butterflies: aerodynamics and mechanical power requirements. J. Exp. Biol. 159:335–57.

————. 1991b. Comparative biomechanics and the evolutionary diversification of flying insect morphology. In: Dudley E. C., ed. The Unity of Evolutionary Biology. Portland: Dioscorides Press. pp. 503–14.

————. 1992. Aerodynamics of flight. In: Biewener A. A., ed. Biomechanics (Structures and Systems): A Practical Approach. Oxford: Oxford Univ. Press. pp. 97–121.

————. 1995. Extraordinary flight performance of orchid bees (Apidae: Euglossini): hovering in heliox (80% He / 20% O2). J. Exp. Biol. 198:1065–70.

————. 1999. The Biomechanics of Insect Flight: Form, Function, Evolution. N.J.: Princeton Univ. Press.

Dudley R., Ellington C. P. 1990a. Mechanics of forward flight in bumblebees. 1. Kinematics and morphology. J. Exp. Biol. 148:19–52.

————. 1990b. Mechanics of forward flight in bumblebees. II. Quasi-steady lift and power requirements. J. Exp. Biol. 148:53–88.

Dudley R., Gans C. 1991. A critique of symmorphosis and optimality models in physiology. Physiol. Zool. 64(3):627–37.

Dudley R., Srygley R. B. 1994. Flight physiology of neotropical butterflies: allometry of airspeeds during natural free flight. J. Exp. Biol. 191:125–39.

Dudley R., Vermeij G. J. 1992. Do the power requirements of flapping flight constrain folivory in flying animals? Function. Ecol. 6:101–4.

————. 1994. Energetic constraints of folivory: leaf fractionation by frugivorous bats. Function. Ecol. 8:668.

Dugard J. J. 1967. Directional change in flying locusts. J. Insect Physiol. 13(7):1055–63.

Dullemeijer P. 1980. Functional morphology and evolutionary biology. Acta Biotheoretica. 29:151–250.

Edmunds G. F., Traver J. R. 1954. The flight mechanics and evolution of the wings of Ephemeroptera, with notes on the archetype insect wing. J. Wash. Acad. Sci. 44(12):390–400.

Edwards R. H., Cheng H. K. 1982. The separation vortex in the Weis-Fogh circulation-generation mechanism. J. Fluid Mech. 120:463–73.

Ellington C. P. 1977. The aerodynamics of normal hovering flight: three approaches. In: Schmidt-Nielsen K., Bolis L., Maddrell S. H. P., Comparative Physiology—Water, Ions, and Fluid Mechanics. London: Cambridge Univ. Press. pp. 327–45.

———. 1980. Vortices and hovering flight. In: Nachtigall W., ed. Instationäre Effekte an Schwingenden Tierflügeln. Wiesbaden: F. Steiner. pp. 64–101.

———. 1984. The aerodynamics of hovering insect flight. Phil. Trans. Roy. Soc. London B 305:1–180.

———. 1995. Unsteady aerodynamics of insect flight. In: Ellington C. P., Pedley T. J., eds. Biological Fluid Dynamics. Vol. 49. Symp. Soc. Exp. Biol. pp. 109–29.

Ellington C. P., Machin K. E., Casey T. M. 1990. Oxygen consumption of bumblebees in forward flight. Nature 347(6292):472–73.

Ellington C. P., van den Berg C., Willmott A. P., Thomas A. L. R. 1996. Leading-edge vortices in insect flight. Nature 384:626–30.

Elton C. S. 1958. The Ecology of Invasions by Animals and Plants. London: Methuen. 181 p.

Emelianov A. F. 1977. Homology of wing structures in Cicadoidea and primitive Polyneoptera. In: Morfologicheskie Osnovy Sistematiki Nasekomykh [Morphological Basis of Insect Taxonomy]. Leningrad: Nauka. pp. 3–48 (in Russian).

Emelianov A. F., Falkovich M. I. 1983. About the book "Historical Development of the Class of Insects." Entomol. Obozr. 62(1):205–22 (in Russian).

———. 1989. A. K. Brodsky. The mechanics of insect flight and the evolution of their wing apparatus. Entomol. Obozr. 68(3):680–82 (in Russian).

Enderlein G. 1903. Die Copeognathen des Indo-Australischen Fannengebietes. Ann. Hist.-Nat. Mus. Nat. Hung. 1:179–344.

Ennos A. R. 1987. A comparative study of the flight mechanism of Diptera. J. Exp. Biol. 127:355–72.

———. 1988. The importance of torsion in the design of insect wings. J. Exp. Biol. 140:137–60.

———. 1989a. The kinematics and aerodynamics of the free flight of some Diptera. J. Exp. Biol. 142:49–85.

———. 1989b. Comparative functional morphology of the wings of Diptera. Zool. J. Linn. Soc. 96:27–47.

———. 1989c. Inertial and aerodynamic torques on the wings of Diptera in flight. J. Exp. Biol. 142:87–95.

———. 1989d. The effect of size on the optimal shapes of gliding insects and seeds. J. Zool. 219:61–69.

Ennos A. R., Wootton R. J. 1989. Functional wing morphology and aerodynamics of *Panorpa germanica* (Insecta: Mecoptera). J. Exp. Biol. 143:267–84.

Esch H., Nachtigall W., Kogge S. N. 1975. Correlations between aerodynamic output, electrical activity in the indirect flight muscles and wing positions of bees flying in a servomechanically controlled wind tunnel. J. Comp. Physiol. 100(2):147–59.

Fabr J.-A. 1914. Instinkt i Nravy Nasekomykh [The Instinct and Habits of Insects]. Vol. 1. Saint Petersburg: Izd. tov-va A. F. Marx. pp. 1–590 (in Russian).

Fabri K. E. 1980. Ethological analysis of the multifunctionality of the chief effector systems of mammals. In: Morfologicheskie Aspekty Evoliutsii [The Morphological Aspects of Evolution]. Moscow: Nauka. pp. 190–210 (in Russian).

Fabricius J. C. 1792. Entomologia systematica emendata et aucta: secundun classes, ordines, genera, species, adjectis synonimis, locis, observationibus, discriptionibus. Table 2. Hafniae: Impensis Christ. Gottl. Proft. 519 p.

Fabrikant N. Y. 1964. Aerodinamika [Aerodynamics]. Moscow: Nauka (in Russian).

Fallén K. F. 1814. Specimen Novam Hemiptera Disponendi Methodum Exhibens. Litteris Berlingianis. Lundae. 26 p.

Favier D., Maresca C., Rebont J. 1982. Dynamic stall due to fluctuations of velocity and incidence. AIAA J. 20:865–71.

Feller P., Nachtigall W. 1989. Flight of the honey bee. II. Inner- and surface thorax temperatures and energetic criteria, correlated to flight parameters. J. Comp. Physiol. B 158:719–27.

Fletcher D. S., Nye J. F. B. 1982. Bombycoidea, Castnioidea, Cossoidea, Mimallonoidea, Sesioidea, Sphingoidea, Zygaenoidea. The Generic Names of Moths of the World. Pt. 4. London. pp. 1–192.

Flower J. W. 1964. On the origin of flight in insects. J. Insect Physiol. 10(1):81–88.

Forbes W. T. M. 1943. The origin of wings and venational types in insects. Am. Midland Natural 29(2):381–405.

———. 1990. Thrust generation by an airfoil in hover modes. Exp. Fluids 9:17–24.

Futuyma D. 1986. Evolutionary Biology. Sunderland, Mass.: Sinauer Associates. 600 p.

Gambarian P. P. 1972. Beg Mlekopitajustchikh [Mammal Running]. Leningrad: Nauka.

Gans C. 1974. Biomechanics. An Approach to Vertebrate Zoology. Ann Arbor: Univ. Michigan Press. 261 p.

———. 1983. On the fallacy of perfection. In: Ray R. R., Gourevich G., eds. Perspectives on Modern Auditory Researches. Groton, Mass.: Amphora pp. 101–14.

———. 1988. Adaptation and the form-function relation. Am. Zool. 28(2):681–97.

———. 1993. On the merits of adequacy. Am. J. Sci. 293-A:391–406.

Garland T., Carter P. A. 1994. Evolutionary physiology. Annu. Rev. Physiol. 56:579–621.

Garland T., Huey R. B. 1987. Testing symmorphosis: does structure match functional requirements? Evolution 41(6):1404–9.

Gettrup E. 1965. Control of forewing twisting by hindwing receptors in flying locusts. In: 12th Int. Congr. Entomol.; 8–16 July 1964; London. pp. 190–92.

Gettrup E., Wilson D. 1964. The lift-control reaction of flying locusts. J. Exp. Biol. 41(1):183–90.

Getzan G., Shimada M., Shimoyama I., Matsumoto Y., Miura H. 1995. Aerodynamic behavior of microstructures. In: Proc. IEEE Symp. on Emerging Technol. and Factory Automation. pp. 54–61.

Gewecke M. 1974. The antennae of insects as air current sense organs and their relationship to the control of flight. Experimental Analysis Insect Behavior. Berlin. pp. 100–13.

———. 1975. The influence of the air-current sense organs on the flight behaviour of *Locusta migratoria*. J. Comp. Physiol. A 103:79–95.

———. 1977. Control of flight in relation to the air in *Locusta migratoria* (Insecta, Orthoptera). J. Physiol. 33:581–92.

———. 1983. Comparative investigations of locust flight in the field and in the laboratory. Vol. 2. In: Nachtigall W., ed. BIONA Rept. pp. 11–20.

Gewecke M., Niehaus M. 1981. Flight and flight control by the antennae in the small tortoiseshell (*Aglais urticae* L., Lepidoptera). 1. Flight balance experiments. J. Comp. Physiol. A 145(2):249–56.

Ghiliarov M. S. 1949. Osobennosti Pochvy kak Sredy Obitania i eio Znachenie v Evoliutsii Nasekomykh [Soil as Environment and Its Role in Evolution of Insects]. Moscow Leningrad: Izd-vo AN SSSR. 280 p.

———. 1957. Evolution of postembryonic development and types of larvae in insects. Zoologicheskij Zhurnal 36(11):1683–97 (in Russian).

Giller P. 1984. Community Structure and the Niche. New York: Chapman & Hall.

Gillott C. 1980. Entomology. New York: Plenum. 729 p.

Goldenberg C. F. 1854. Die fossilen Insekten der Kohlenformation von Saarbrücken. Paleontographica 4:17–40.

Goodwin B. 1976. Analytical Physiology of Cells and Developing Organisms. New York: Academic.

Gorodkov K. B. 1984. Oligomerisation and the evolution of systems of morphologic structures. 2. Oligomerisation and the decrease of body size. Zool. Zhurn. 63(12): 1765–78 (in Russian).

Götz K. G. 1968. Flight control in *Drosophila* by visual perception of motion. Kybernetik 4(6):199–208.

———. 1987. Course-control, metabolism and wing interference during ultralong tethered flight in *Drosophila melanogaster*. J. Exp. Biol. 128:35–46.

Götz K. G., Wandel U. 1984. Optomotor control of the force of flight in *Drosophila* and *Musca*. 2. Covariance of lift and thrust in still air. Biol. Cybern. 51(2):135–39.

Gould S. J. 1966. Allometry and size in ontogeny and phylogeny. Biol. Rev. 42:587–640.

———. 1971. Geometric similarity in allometric growth: a contribution to the problem of scaling in the evolution of size. Am. Natural 105(942):113–36.

———. 1974. The origin and function of "bizarre" structures: antler size and skull size in the "Irish Elk," *Megaloceros giganteus*. Evolution 28:191–220.

———. 1977. Ontogeny and Phylogeny. Cambridge, Mass.: Harvard Univ. Press. 501 p.

———. 1984. Morphological channeling by structural constraint: convergence in styles of dwarfing and gigantism in *Cerion*, with a description of two new fossil species and a report on the discovery of the largest *Cerion*. Paleobiology 10(2):172–94.

————. 1989. A developmental constraint in *Cerion*, with comments on the definition and interpretation of constraint in evolution. Evolution 43(3):516–39.

————. 1992. Constraint and the square snail: life and the limits of a covariance set. The normal teratology of *Cerion disforme*. Biol. J. Linnean Soc. 47:407–37.

Gould S. J., Eldredge N. 1977. Punctuated equilibria: the tempo and mode of evolution reconsidered. Paleobiology 3:115–51.

————. 1993. Punctuated equilibrium comes of age. Nature 366(6452):223–27.

Govind C. K., Burton A. J. 1970. Flight orientation in a coreid squash bug (Heteroptera). Can. Entomol. 102(8):1002–7.

Graham J. B., Aguilar N., Dudley R., Gans C. 1997. The late Paleozoic atmosphere and the ecological and the evolutionary physiology of tetrapods. In: Amniote Origins. New York: Academic. pp. 141–66.

Graham J. B., Dudley R., Aguilar N. M., Gans C. 1995. Implications of the late Palaeozoic oxygen pulse for physiology and evolution. Nature 375:117–20.

Gratshev V. G., Zherikhin V. V. 1993. New fossil mantids (Insecta, Mantida). Paleontol. J. 27(1A):148–65.

Greenewalt C. H. 1960. The wings of insects and birds as mechanical oscillators. Proc. Am. Phil. Soc. 104(6):605–11.

Gringorten J. L., Friend W. G. 1979. Wing beat pattern in *Rhodnius prolixus* Stal (Heteroptera: Reduviidae) during exhaustive flight. Can. J. Zool. 57(2):391–95.

Grodnitsky D., Dudley R. 1996. Vortex visualization during free flight of Heliconiine butterflies (Lepidoptera: Nymphalidae). J. Kansas Entomol. Soc. 69(2):199–203.

Grodnitsky D., Dudley R., Gilbert L. 1994. Wing decoupling in hovering flight of Papilionid butterflies. Tropic. Lepidoptera 5(2):85–86.

Grodnitsky D. L. 1988. The structure and possible functions of the wing scale covering of butterflies (Lepidoptera, Hesperioidea, Papilionoidea). Entomol. Obozr. 67(2): 251–56 (in Russian, published in English in Entomol. Rev. [Wash.] 1989; 68(2):11–17).

————. 1991a. On the wing venation of caddis flies, moths and butterflies (Phryganeida, Papilionida). Zool. Zhurn. 70(6):77–87 (in Russian).

————. 1991b. Folding of the wings in caddisflies, moths and butterflies (Insecta: Phryganeida, Papilionida). Vestn. Zool., no. 5:34–41 (in Russian).

————. 1992a. Free and tethered flight of butterflies (Papilionida: Papilionoidea). Zool. Zhurn. 71(4):21–28 (in Russian).

————. 1992b. The influence of experimental conditions on the lift generated by tethered flying insects. Siberian Biol. Zhurn., no. 6:46–49 (in Russian).

————. 1993. Preliminary data on the body motion of butterflies during free flight. Zool. Zhurn. 72(7):84–94 (in Russian).

————. 1994. The evolution of insect flight. Priroda, no. 8:27–32 (in Russian).

————. 1995a. Classification and evolution of insect flight kinematics. Evolution 49(6): 1158–62.

————. 1995b. Problems of functional interpretation of some similar morphologic structures on insect wings and the explanation of the secondary similarity of organisms. Zhurn. Obstchej Biologii 56(4):438–49 (in Russian).

————. 1996. Adaptations for flapping flight in different endopterygote insects. Zool. Zhurn. 75(5):692–700 (in Russian).

Grodnitsky D. L., Kozlov M. V. 1985. The functional morphology of wing apparatus and peculiarities of flight of primary moths (Lepidoptera, Micropterigidae, Eriocraniidae). Zool. Zhurn. 64(11):1661–71 (in Russian).

————. 1987. The dependence of codling moth *Laspeyresia pomonella* L. (Lepidoptera, Tortricidae) tethered flight kinematics on the conditions of experiment. Zool. Zhurn. 66(9):1314–20 (in Russian).

————. 1989a. The structural and functional organization of wing scale covering of butterflies and moths (Papilionida=Lepidoptera). Uspekhi Sovremennoj Biologii 107(3):457–68 (in Russian).

————. 1989b. Morphology of wing scale covering in swallowtail butterflies (Lepidoptera, Papilionidae). Zool. Zhurn. 68(3):41–49 (in Russian).

————. 1990a. Morphology and evolution of wing scale covering in lappet moths (Papilionida: Lasiocampidae). Zool. Zhurn. 69(4):44–51 (in Russian, published in English in Entomol. Rev. [Wash.] 1990; 69(5):145–153).

————. 1990b. Wing functional morphology of some species of Papilionina (Lepidoptera) suborder. Vestn. Zool., no. 2:58–64 (in Russian).

————. 1991. Evolution and functions of wings and their scale covering in butterflies and moths (Insecta: Papilionida = Lepidoptera). Biologisches Zentralblatt 110(3): 199–206.

Grodnitsky D. L., Kozlov M. V., Nesina M. V. 1988. The problem of elaboration of a well-fitted model of insect wing apparatus. 1. Literature review. Uspekhi Sovremennoj Biologii 105(2):284–99 (in Russian).

Grodnitsky D. L., Morozov P. P. 1992. Flow visualization experiments on tethered flying green lacewings *Chrysopa dasyptera*. J. Exp. Biol. 169:143–63.

————. 1993. Vortex formation during tethered flight of functionally and morphologically two-winged insects, including evolutionary considerations on insect flight. J. Exp. Biol. 182:11–40.

————. 1994. Morphology, flight kinematics and deformation of the wings in holometabolous insects (Insecta: Oligoneoptera = Scarabaeiformes). Russian Entomol. J. 3(3/4):3–32.

————. 1995. The vortex wake of flying beetle. Zool. Zhurn. 74(3):66–72 (in Russian).

Gurgey E., Thiele F. 1991. Numerical simulation of the viscous flow over an oscillating airfoil. In: Khalighi B., Braun M. T., Freitas C. T., eds. Experimental and Numerical Flow Visualization. New York: American Soc. of Mechanical Engineers. pp. 377–83.

Gursul I., Ho C.-M. 1992. High aerodynamic loads on an airfoil submerged in an unsteady stream. AIAA J. 30:1117–19.

Haeckel E. H. P. A. 1896. Systematische Phylogenie der wirbellosen Thieren (Invertebrata). Berlin: G. Reimer. 720 p.

Haliday A. H. 1836. An epitome of the British genera, in the order Thysanoptera, with indications of a new species. Entomol. Mag. 3:439–51.

Hamilton K. G. A. 1971. The insect wing. Part 1. Origin and development of wings from notal lobes. J. Kansas Entomol. Soc. 44(4):421–33.

———. 1972a. The insect wing. 2. Vein homology and archetypal insect wing. J. Kansas Entomol. Soc. 45(1):54–58.

———. 1972b. The insect wing. 3. Venation of the orders. J. Kansas Entomol. Soc. 45(2):145–62.

Handlirsch A. 1903. Zur Phylogenie der Hexapoden. Sitzungsberichte der (K.) Akademie der Wissenschaften in Wien, mathematische-naturwissenschaftliche klasse. sitz 112(1):716–38.

———. (1906–1908). Die fossilen Insekten und die Phylogenie der rezenten Formen. Leipzig: Wilhelm Engelmann. 1430 p.

———. 1911. New Paleozoic insects from the vicinity of Mason Creek, Illinois. Am. J. Sci. (ser. 4). 31:297–326, 353–77.

———. 1925. Systematische Übersicht. In: Schröder C., ed. Handbuch der Entomologie. Vol. 3. Jena: Gustav Fischer. pp. 377–1140.

———. 1937. Neue Untersuchungen über die fossilen Insekten. Part 1. Annalen des naturhistorischen Museums in Wien. 48:1–140.

Handschin E. 1958. Die systematische Stellung der Collembolen. In: Proc. 10th Int. Congr. Entomol.; 1956; Montreal. Vol. 1, pp. 499–508.

Hanegan J. L., Heath J. E. 1970a. Mechanisms for the control of body temperature in the moth, *Hyalophora cecropia*. J. Exp. Biol. 53(2):349–62.

———. 1970b. Temperature dependence of the neural control of the moth flight system. J. Exp. Biol. 53(3):629–39.

Heath J. E., Adams P. A. 1965. Temperature regulation in the sphinx moth during flight. Nature 205(4968):309–10.

Heinrich B. 1970. Thoracic temperature stabilization by blood circulation in a free-flying moth. Science 168(3931):580–82.

———. 1971. Temperature regulation of the sphinx moth, *Manduca sexta*. 1. Flight energetics and body temperature during free and tethered flight. J. Exp. Biol. 54(1):141–52.

———. 1974. Thermoregulation in endothermic insects. Science 185(4153):747–56.

———. 1981. Ecological and evolutionary perspectives. In: Heinrich B., ed. Insect Thermoregulation. New York: Wiley. pp. 235–302.

———. 1993. The hot-blooded insects. Strategies and mechanisms of thermoregulation. Cambridge, Mass.: Harvard Univ. Press. 601 p.

Heinrich B., Bartholomew G. A. 1971. An analysis of pre-flight warm-up in the sphinx moth *Manduca sexta*. J. Exp. Biol. 55(1):223–39.

———. 1972. Temperature control in flying moths. Sci. Am. 226(6):71–77.

Heinrich B., Mommsen T. P. 1985. Flight of winter moths near 0°C. Science 228(4696): 177–79.

Heisenberg M., Wolf R. 1979. On the fine structure of yaw torque in visual flight orientation of *Drosophila melanogaster*. J. Comp. Physiol. A 130:113–30.

Heisenberg V. 1969. Der Teil und das Ganze. München.

Hennig W. 1981. Insect Phylogeny. Chichester: Wiley pp. 1–514.

Hewson R. J. 1969. Some observations on flight in *Oncopeltus fasciatus* (Hemiptera: Lygaeidae). J. Entomol. Soc. Brit. Columbia 66:45–49.

Hinton H. E. 1963. The origin of flight in insects. Proc. Roy. Entomol. Soc. London (C) 28:24–25.

Hisada M., Tamasige M., Suzuki N. 1965. Control of the dragonfly *Sympetrum darwinianum* Selys. 1. Dorsophotic response. J. Fac. Sci. Hokkaido Univ. (ser. 6, Zool.) 15:568–77.

Hochachka P. W., Somero G. N. 1973. Strategies of Biochemical Adaptation. Philadelphia, W. B. Saunders.

Hollick F. S. J. 1940. The flight of the dipterous fly *Muscina stabulans* Fallen. Phil. Trans. Roy. Soc. London B 230:357–90.

Hummel D., Goslow G. E. 1991. Concluding remarks: bird flight. In: Acta 20th Congr. Int. Ornithol. Wellington. p. 748.

Hutchinson G. E. 1959. Homage to Santa Rosalia, or why are there so many kinds of animals? Am. Natural 93(868):145–59.

Iljichev V. D. 1973. Adaptations—ecological parallelisms—mozaic evolution (bird acoustic system as an object of functional morphology). Zhurn. Obstchej Biologii 34(1):66–80 (in Russian).

Iordansky N. N. 1994. Makroevoliutsia. Sistemnaja Teorija. [Macroevolution. A Systemic Theory]. Moscow: Nauka. 113 p. (in Russian).

Ivanov V. D. 1985a. Comparative analysis of wingbeat kinematics of caddisflies (Trichoptera). Entomol. Obozr. 64(2):273–84 (in Russian).

———. 1985b. Morphology and evolution of wing articulation of caddisflies. 1. The initial type of construction. Vestn. LGU, no. 10:3–12 (in Russian).

———. (1987a.) Morphology and evolution of wing articulation of caddisflies. 2. The wing articulation of Integripalpia. Vestn. LGU, no. 3:11–21 (in Russian).

———. (1987b). Morphology and evolution of wing articulation of caddisflies. 3. The wing articulation of Annulipalpia. Vestn. LGU, no. 17:15–25 (in Russian).

———. 1989. Evolution of flight of caddis flies. In: Tomaszewski C., ed. Proc. 6th Int. Symp. on Trichoptera. Lodz-Zakopane. 12–16 Sept. 1989. Adam Mickiewicz Univ. Press. pp. 351–57.

———. 1990. Comparative analysis of flight aerodynamics of caddisflies (Insecta: Trichoptera). Zool. Zhurn. 69(2):46–60 (in Russian).

———. 1994. Comparative analysis of the construction of wing articulations of archaic lepidopterans. Entomol. Obozr. 73(3):569–90 (in Russian).

Ivlev Yu. F. 1993a. Heat insulatory properties and structure of covering in endothermic lepidopterans of Noctuidae family. 1. Heat insulation. Zool. Zhurn. 72(12):25–39 (in Russian).

———. 1993b. Heat insulatory properties and structure of covering in endothermic lepidopterans of Noctuidae family. 2. Structure. Zool. Zhurn. 72(12):40–50 (in Russian).

Jean R. V. 1978. Growth and entropy: Phylogenism and phyllotaxis. J. Theor. Biol. 71(4):639–60.

Jedlička L. 1986. World distribution of mountain midges (Deuterophlebiidae). In: Abstr. 1st Int. Congr. Dipterol., Budapest. 17–24 Aug. 1986. Budapest. p. 111.

Jedlička L., Halgoš J. 1981. *Deuterophlebia sajanica* sp. n., a new species of mountain midges from Mongolia (Diptera: Deuterophlebiidae). Biologia (Bratislava) 36(11): 973–81.

Jensen M. 1956. Biology and physics of locust flight. 3. The aerodynamics of locust flight. Phil. Trans. Roy. Soc. London B 239:511–52.

Jungmann R., Rothe U., Nachtigall W. 1989. Flight of the honey bee. 1. Thorax surface temperature and thermoregulation during tethered flight. J. Comp. Physiol. B 158:711–18.

Kaganova Z. V. 1972. The concept of structural levels and integrative principles in contemporary biology. In: Razvitie Kontseptsii Strukturnykh Ourovnej v Biologii [Development of the Concept of Structural Levels in Biology]. Moscow: Nauka. pp. 112–21 (in Russian).

Kammer A. E. 1971. The motor output during turning flight in a hawkmoth, *Manduca sexta*. J. Insect Physiol. 17(6):1073–86.

———. 1985. Flying. In: Kerkut G. A., Gilbert L. I., eds. Comprehensive Insect Physiology, Biochemistry and Pharmacology. Vol. 5. Oxford: Pergamon. pp. 491–552.

Kayser H. 1985. Pigments. In: Kerkut G. A., Gilbert L. I., eds. Comprehensive Insect Physiology, Biochemistry and Pharmacology. Vol. 10. Oxford: Pergamon. pp. 367–416.

Kettlewell H. B. D. 1955. Selection experiments on industrial melanism in the Lepidoptera. Heredity 10:287–301.

Kingsolver J. G., Koehl M. A. R. 1989. Selective factors in the evolution of insect wings: response to Kukalova-Peck. Can. J. Zool. 67(3):785.

———. 1994. Selective factors in the evolution of insect wings. Annu. Rev. Entomol. 39:425–51.

Kingsolver J., Koehl M. A. R. 1985. Aerodynamics, thermoregulation, and the evolution of insect wings: differential scaling and evolutionary change. Evolution 39(3):488–504.

Kirby W. 1813. Strepsiptera: a new order of insects proposed; and the characters of the order, with those of its genera laid down. Trans. Linnean Soc. London 11:(Pt. 1)86–123.

———. 1826. An Introduction to Entomology, or, Elements of the Natural History of Insects: With Plates. Vol. 3. London: Longman. 732 p.

Kliuge N. Yu. 1989. The problem of homology of branchial gills and paranotal lobes in mayfly nymphs to insect wings as related to taxonomy and phylogeny of the mayfly order (Ephemeroptera). In: Chtenia Pamiati N. A. Kholodkovskogo. Dokl. na 41om Ezhegod. Chtenii April 1 1988. Leningrad: Nauka. pp. 48–77 (in Russian).

Kofman G. B. 1981a. The biological meaning of allometric regularities. In: Terskov I. A., ed. Issledovanija Dinamiki Rosta Organizmov [Investigation of the Growth Dynamics in Organisms]. Novosibirsk: Nauka. pp. 36–55 (in Russian).

———. 1981b. The methods of similarity and dimensionalities in the studies of relative growth of organisms. Zhurn. Obstchej Biologii 42(2):234–40 (in Russian).

———. 1986. Rost i Forma Dereviev [Growth and Form of Trees]. Novosibirsk: Nauka. 208 p. (in Russian).

Kokshaysky N. V. 1970. Energetics of flight of insects and birds. Zhurn. Obstchej Biol. 31(5):527–49 (in Russian).

———. 1973. Functional aspects of some details of bird wing configuration. Syst. Zool. 22(4):442–50.

———. 1974. Ocherki Biologicheskoj Aero- i Gidrodinamiki [Essays on Biological Aero- and Hydrodynamics]. Moscow: Nauka. 254 p. (in Russian).

———. 1979. Tracing the wake of a flying bird. Nature 279(5709):146–48.

———. 1980. On the relationships between form and function and their transformation in phylogeny. In: Morfologicheskie Problemy Evoliutsii [Morphological Aspects of Evolution]. Moscow: Nauka. pp. 37–53 (in Russian).

———. 1982. The contribution of domestic science to the studies of bird flight. Zool. Zhurn. 61(7):971–87 (in Russian).

———. 1988a. On the principle of the evolutionary stabilization of functions. In: Sovremennyje Problemy Evoliutsionnoj Morfologii [Modern Problems of Evolutionary Morphology]. Moscow: Nauka. pp. 28–47 (in Russian).

———. 1988b. The principle of evolutionary stabilization of functions in the behavior of animals. Zool. Zhurn. 67(2):176–88 (in Russian).

Kokshaysky N. V., Petrovsky V. I. 1979a. Preliminary data on the nature of wake of flying bird. Dokl. AN SSSR 244(5):1248–51 (in Russian).

———. 1979b. The wake of flying bird. Priroda, no. 5:100–102 (in Russian).

Kolmogorov A. N. 1991. Matematika v eje Istoricheskom Razvitii [Mathematics in Its Historical Development]. Moscow: Nauka. 224 p. (in Russian).

Komarov V. T., Mordvinov I. E. 1989. Kinematic peculiarities of coot (Fulica atra) flight as related to its ecological morphology. Zool. Zhurn. 68(6):93–98 (in Russian).

Korochkin L. I. 1986. Molecular and genetical aspects of ontogeny. In: Biologija Razvitija i Upravlenie Nasledstvennostiju [Developmental Biology and the Control over Heredity]. Moscow: Nauka. pp. 267–84 (in Russian).

Koucherov I. B. 1995. On the principle of complementarity in geobotany: the methodological prerequisites of the appearance of additive approaches to the study of vegetation. Zhurn. Obstchej Biologii 56(4):486–505 (in Russian).

Koussakin O. G., Drozdov A. L. 1994. Filema Organicheskogo Mira. Ch. 1. Prolegomeny k Postroeniju Sistemy [The Organic World Phylogeny. Part 1. Prolegomena for the Phylogeny Construction]. Saint Petersburg: Nauka. 284 p. (in Russian).

Kouzin B. S. 1987. The principles of taxonomy. In: Voprosy Istorii Estestvoznanija i Tekhniki [The Queries of Natural History and Technics]. Moscow: Nauka. no. 4, pp. 137–42 (in Russian).

Kozhov M. 1963. Lake Baikal and Its Life. Hague: Dr. W. Junk, Publishers. 345 p.

Kozhov M. M. 1973. The origin and evolutionary pathways of the Baikal lake fauna. In: Problemy Evoliutsii [Problems of Evolution]. Vol. 3. Novosibirsk: Nauka. pp. 5–30 (in Russian).

Kozlov L. F. 1983. Teoreticheskaja Biogidrodinamika [Theoretical Biological Fluid Dynamics]. Kiev: Vischa shkola. 240 p. (in Russian).

Kozlov M. V. 1987. Functional morphology of the wings and variability of their venation in primitive lepidopterans (Lepidoptera: Micropterigidae, Tischeriidae). Zhurn. Obstchej Biol. 48(2):238–47 (in Russian).

———. 1988. Lepidopteran paleontology and problems of the phylogeny of Papilionida order. In: Melovoj Biotsenoticheskij Krisis I Evoliutsia Nasekomykh [The Cretaceous Biocoenotic Crisis and the Evolution of Insects]. Moscow: Nauka. pp. 16–69 (in Russian).

Kozlov M. V., Ivanov V. D., Grodnitsky D. L. 1986. The evolution of wing apparatus and wingbeat kinematics in lepidopterans. Uspekhi Sovremennoj Biologii 101(2):291–305 (in Russian).

Kramer M. G., Marden J. H. 1997. Almost airborne. Nature 385(6615):403–4.

Krassilov V. A. 1986. Nereshennye Problemy Teorii Evoliutsii [Unsolved Problems of the Theory of Evolution]. Vladivostok: BPI DVO AN SSSR 140 p. (in Russian).

Krenke N. 1927. Rules of leaf shape combination at opposite and alternate arrangement. In: Trudy po Prikladnoj Botanike i Selektsii. [Studies on Applied Botany and Selection] Moscow: NII im. K.A. Timiriazeva. Vol. 17, no. 2, pp. 71–168 (in Russian).

Kreslavsky A. G. 1991. On the origin of morphological innovations. In: Sovremennaja Evoliutsionnaja Morfologia [Contemporary Evolutionary Morphology]. Kiev: Naukova Doumka. pp. 176–89 (in Russian).

Kristensen N. P. 1970. Morphological observations on the wing scales in some primitive Lepidoptera (Insecta). J. Ultrastructure Res. 30(3–4):402–10.

———. 1974. On the evolution of wing transparency in Sesiidae (Insecta, Lepidoptera). Vidensch. Medd. Dansk. Naturhist. foren. 137:125–34.

———. 1978. Ridge dimorphism and second-order ridges on wing scales in Lepidoptera: Exoporia. Int. J. Insect Morphol., Embryol. 7(3):297–99.

———. 1981a. Revisionary note no. 336. In: Hennig W. 1981. Insect Phylogeny. Chichester: Wiley pp. 1–514.

———. 1981b. Phylogeny of insect orders. Ann. Rev. Entomol. 26:135–57.

———. 1984. Studies on the morphology and systematics of primitive Lepidoptera (Insecta). Steenstrupia 10(5):141–91.

———. 1992. Phylogeny of extant Hexapods. In: The Insects of Australia. 2nd ed. Melbourne: Melbourne Univ. Press. pp. 125–40.

Kristensen N. P., Nielsen E. S. 1979. A new subfamily of micropterigid moths from South America. A contribution to the morphology and phylogeny of the Micropterigidae, with a generic catalogue of the family (Lepidoptera: Zeugloptera). Steenstrupia 5(7):69–147.

Kukalova J. 1958. Paoliidae Handlirsch (Insecta—Protorthoptera) aus dem Oberschlesischen Steinkohlenbecken. Geologie 7:935–59.

Kukalova-Peck J. 1978. Origin and evolution of insect wings, and their relation to metamorphosis, as documented by the fossil record. J. Morphol. 156:53–125.

———. 1983. Origin of the insect wing and wing articulation from the arthropodan leg. Can. J. Zool. 61(7):1618–69.

———. 1985. Ephemeroid wing venation based upon new gigantic Carboniferous mayflies and basic morphology, phylogeny, and metamorphosis of pterygote insects (Insecta, Ephemerida). Can. J. Zool. 63:933–55.

————. 1987. New carboniferous Diplura, Monura, and Thysanura, the hexapod ground plan, and the role of thoracic side lobes in the origin of wings (Insecta). Can. J. Zool. 65:2327–45.

————. 1991. Fossil history and the evolution of hexapod structures. The Insects of Australia. Vol. 1. Melbourne: Melbourne Univ. Press. pp. 141–79.

————. 1992. Fossil history and the evolution of hexapod structure. In: The Insects of Australia. 2nd ed. Melbourne: Melbourne Univ. Press. pp. 144–82.

————. 1992. The "Uniramia" do not exist: the ground plan of the Pterygota as revealed by permian Diaphanopterodea from Russia (Insecta: Palaeodictyoptera). Can. J. Zool. 70(2):236–55.

Kukalova-Peck J., Lawrence J. F. 1993. Evolution of the hind wing in Coleoptera. Can. Entomol. 125:181–258.

Kutsch W., Gewecke M. 1981. Development of flight behavior in maturing adults of *Locusta migratoria*. 2. Aerodynamic parameters. J. Insect Physiol. 27:455–59.

Kutsch W., Schwarz G., Fischer H., Kautz H. 1993. Wireless transmission of muscle potentials during free flight of a locust. J. Exp. Biol. 185:367–73.

Kutsch W., Stevenson P. 1981. Time-correlated flights of juvenile and mature locusts: a comparison between free and tethered animals. J. Insect Physiol. 27:455–59.

Labandeira, C. C., Beall B. S., Hueber F. M. 1988. Early insect diversification: evidence from Lower Devonian bristletail from Quebec. Science 242:913–16.

LaBarbera M. 1990. Principles of design of fluid transport systems in zoology. Science 249(4972):992–1000.

Lampert K. 1913. The atlas of lepidopterans and caterpillars of Europe and partly Russian-Asian territories. Saint Petersburg: Izd. A. F. Devrien. 486 p. (in Russian).

Lan C. E. 1979. The unsteady quasi-vortex-lattice method with applications to animal propulsion. J. Fluid Mech. 93:747–65.

Lande R. 1978. Evolutionary mechanisms of limb loss in tetrapods. Evolution 32:73–92.

Lang A. 1888. Lehrbuch der vergleichenden Anatomie der wirbellosen Thiere. Vol. 1. Jena, Germany: Fischer.

Latreille P. A. 1796. Précis des caractères génériques des insectes, disposés dans un ordre naturel. Brive, France: F. Bourdeaux. 210 p.

————. 1802. Histoire naturelle, générale et particulière de crustacés et insectes. Table 3. Paris: Dufart. 467 p.

————. 1804. Histoire naturelle, générale et particulière, des crustacés et des insectes. Vol. 13. Paris. 458 p.

————. 1806. Genera Crustaceorum et Insectorum. Vol. 1. Paris: A. Koenig. 302 p.

————. 1810. Considérations genérales sur l'ordre naturelles animaux composant les classes des crustacés, des arachnides, et des insectes. Paris: Schoell. 444 p.

————. 1825. Familles naturelles de régne animal. Paris: Baillière. 570 p.

Lauder G. V. 1990. Functional morphology and systematics: studying functional patterns in an historical context. Annu. Rev. Ecol. Systemat. 21:317–40.

Leach W. E. 1815. Artikel entomology. In: Brewster, ed. Edinburgh Encyclopaedie. Vol. 9. Edinburgh. pp. 57–172.

Leech R., Cady A. 1994. Function shift and the origin of insect flight. Australian Biologist 7(4):160–68.

Lehmann F.-O. 1994. Aerodynamische, kinematische und electrophysiologische Aspekte der Flugkrafterzeugung und Flugkraftstenerung bei der Taufliege *Drosophila melanogaster*. Ph.D. diss., Univ. of Tübingen, Germany.

Lehmann F.-O., Dickinson M. H. 1997. The changes in power requirements and muscle efficiency during elevated force production in the fruit fly, *Drosophila melanogaster*. J. Exp. Biol. 200:1133–1143.

Liem K. F. 1988. Form and function of lungs: the evolution of air breathing mechanisms. Am. Zool. 28:739–59.

Lighthill M. J. 1969. Hydromechanics of aquatic animal propulsion—a survey. Annu. Rev. Fluid Mech. 1:413–46.

———. 1973. On the Weis-Fogh mechanism of lift generation. J. Fluid Mech. 160:1–17.

———. 1978. A note on "clap and fling" aerodynamics. J. Exp. Biol. 173:279–80.

Lima de Faria A. 1988. Evoliutsija bez Otbora [Evolution without Selection]. Moscow: Mir (in Russian).

Linné C. 1758. Systema naturae per regnum tria naturae, secundum classes, ordina, genera, species, cum characteribus, differentiis, synonymis, locis, ed. X. Laur. Salvii: Holmiae. 1767 p.

Liubarsky G. Yu 1991a. "And so the entire choir points at a mysterious law . . . ." Znanie-Sila no. 10:34–41 (in Russian).

———. 1991b. Objectification of the category of taxonomy rank. Zhurn. Obstchej Biologii 52(5):613–26 (in Russian).

———. 1993. The method of general typology in biological research. 1. Comparative method. Zhurn. Obstchej Biologii 54(4):408–29 (in Russian).

———. 1994. [Review of the book by Levchenko V. F.: Models in the Theory of Biological Evolution. SPb: Nauka, 1993. 382 p.]. Zhurn. Obstchej Biologii 55(3):375–82 (in Russian).

Lobashev M. E. 1963. Genetika [Genetics]. Leningrad: Izd-vo LGU. 490 p. (in Russian).

Lubischew A. A. 1982. Problemy Formy, Sistematiki I Evoliutsii Organizmov [Problems of Organismal Form, Taxonomy and Evolution]. Moscow: Nauka. 280 p. (in Russian).

Magnan A. 1934. La locomotion chez les animaux. 1. Le vol des insectes. Paris: Hermann & Co. 186 p.

Maiorana V. C. 1978. An explanation of ecological and developmental constants. Nature 273(5661):375–77.

Malakhov V. V. 1980. Organism from the viewpoint of a morphologist. In: Urovni Organizatsii Biologicheskikh Sistem [The Levels of Organization of Biological Systems]. Moscow: Nauka. pp. 76–96 (in Russian).

Mamaev B. M. 1975. The gravitational hypothesis of the origin of insects. Entomol. Obozr. 54(3):449–506 (in Russian, published in English in Entomol. Rev. [Wash.] 1975; 54(3):13–17).

Mamkaev Yu. V. 1983. On the value of ideas by V. A. Dogel' for the evolutionary morphology. In: Evoliutsionnaja Morfologija Bespozvonochnykh [The Evolutionary Morphology of Invertebrates]. Leningrad: Nauka (Proc. Zool. Inst. AN SSSR. Vol. 109). pp. 15–36 (in Russian).

———. 1984. Morphological irradiation and the problem of parallelisms. In: Makroevoliutsia [Macroevolution]. Moscow: Nauka. pp. 85–87 (in Russian).

———. 1991. The methods and regularities of evolutionary morphology. In: Sovremennaja Evoliutsionnaja Morfologija [Contemporary Evolutionary Morphology]. Kiev: Naukova Doumka. pp. 33–56 (in Russian).

Marden J. H. 1987. Maximum lift production during takeoff in flying animals. J. Exp. Biol. 130:235–58.

Marden J., Kramer M. 1994. Surface-skimming stoneflies: a possible intermediate stage in insect flight evolution. Science 266:427–30.

———. 1995. Locomotor performance of insects with rudimentary wings. Nature 377(6547):332–34.

Mareš J., Lapàček V. 1980. Nejkrasnejši brouci tropů. Prague: Academia. 108 p.

Maresca C., Favier D., Rebont J. 1979. Experiments on an aerofoil at high angle of incidence in longitudinal oscillations. J. Fluid Mech. 92:671–90.

Margalef R. 1992. A View of the Biosphere. Moscow: Nauka. 214 p. (in Russian).

Martin L., Carpenter P. 1977a. Flow-visualization experiments on butterflies in simulated gliding flight. In: Physiology of movement, biomechanics. Symposium Mainz. Oct. 1976. Vol. 1. Stuttgart: Fischer. pp. 307–16.

———. 1977b. Flow-visualization experiments on butterflies in simulated gliding flight. Fortschr. Zool. 24:307–16.

Martynov A. V. 1923. On the two basic wing types and their meaning for the general classification of insects. In: Trans. 1st All-Russian Meeting of Zoologists, Anatomists and Hystologists, Petrograd. 15–22 Dec. 1922. Petrograd. pp. 88–89 (in Russian).

———. 1924a. Sur l'interpretation de la nervuration et de la tracheation des ailes des Odonates et Agnathes. Russ. Entomol. Obozr. 18:145–74 (in Russian, published in English in Psyche 1930; 37:245–80).

———. 1924b. On the two types of insect wings and their evolution. Russ. Zool. Zhurn. 4:155–85 (in Russian).

———. 1925. Ueber zwei Grundtypen der Flügel bei den Insekten und ihre Evolution. Ztschr. Oekol. Morph. der Tiere 4(3):465–501.

———. 1927. Über eine neue Ordnung der fossilen Insekten, Miomoptera nov. Zool. Anz. 72:99–109.

———. 1928. Permian fossil insects of North-East Europe. In: Trudy geol. Mus. Akad. Nauk SSSR. Vol. 4, pp. 1–118.

———. 1932. New Permian Palaeoptera with the discussion of some problems of their evolution. In: Trudy Paleozool. Inst. Akad. Nauk SSSR. Moscow: Paleozool. Inst. AN SSSR. Vol. 1. pp. 1–44.

———. 1938. Essays on the geological history and phylogeny of the insect orders (Pterygota). Pt. I. Palaeoptera and Neoptera-Polyneoptera. In: Trudy Paleontologicheskogo Inst. Akad. Nauk SSSR. Moscow: Izd-vo AN SSSR. Vol. 7, no. 4. 148 p. (in Russian, with French summary).

Martynova O. M. 1960. On wing venation in lepidopterans (Lepidoptera). Entomol. Obozr. 39(2):296–99 (in Russian, published in English in Entomol. Rev. [Wash.] 1960; 39(2):190–91).

Maslov S. P. 1980. The limitations of the opportunities of homeostasis by multifunctionality and the chief ways of its bypass. In: Urovni Organizatsii Biologicheskikh Sistem [The Levels of Biological Organization]. Moscow: Nauka. pp. 8–19 (in Russian).

———. 1984. Multifunctionality and multivariate support of functions: relationships and the role in evolution. In: Makroevoliutsija. Proc. 1st All-Union. Conf. on the Problems of Evolution. Moscow. pp. 232–33 (in Russian).

Mason W. R. M. 1986. Standard drawing conventions and definitions for venation and other features of wing of Hymenoptera. Proc. Entomol. Soc. Wash. 88:1–7.

Matienko B. T. 1981. The principles of evolution and adaptive transformations of plant structure and ultrastructure. Izv. AN MSSR. Ser. Biol. Chem. Sci., no. 3:5–27 (in Russian).

Matveev B. S. 1945. On the system of correlated changes of shape, functions and environment in the evolution of animals. Zool. Zhurn. 24(1):3–22 (in Russian).

———. 1957. On the transformation of function in the individual development of animals. Zool. Zhurn. 36(1):4–25 (in Russian).

Maxworthy T. 1972. The structure and stability of vortex rings. J. Fluid Mech. 51(Pt. 1):15–32.

———. 1979. Experiments on the Weis-Fogh mechanism of lift generation by insects in hovering flight. Part 1. Dynamics of the "fling." J. Fluid Mech. 93:47–63.

———. 1981. The fluid dynamics of insect flight. Annu. Rev. Fluid Mech. 113:329–50.

May M. L. 1981. Wingstroke frequency of dragonflies (Odonata: Anisoptera) in relation to temperature and body size. J. Comp. Physiol. 144(2):229–40.

———. 1991. Dragonfly flight: power requirements at high speed and acceleration. J. Exp. Biol. 158:325–42.

May M. L., Brodfuehrer P. D., Hoy R. R. 1988. Kinematic and aerodynamic aspects of ultrasound-induced negative phonotaxis in flying Australian field crickets (*Teleogryllus oceanicus*). J. Comp. Physiol. A 164(2):243–49.

May M., Hoy R. R. 1990. Leg-induced steering in flying crickets. J. Exp. Biol. 151:485–88.

Maynard Smith J. 1978. The Evolution of Sex. Cambridge: Cambridge Univ. Press.

Mayr E. 1969. Principles of Systematic Zoology. New York: McGraw-Hill.

———. 1970. Populations, Species and Evolution. Cambridge, Mass.: The Belknap Press of Harvard Univ. Press.

McCroskey W. J. 1982. Unsteady airfoils. Annu. Rev. Fluid Mech. 14:285–311.

McGahan J. 1973. Flapping flight of the Andean Condor. J. Exp. Biol. 58(1):239–53.

McMahon T. 1973. Size and shape in biology. Science 179(4079):1201–4.

———. 1975. Allometry and biomechanics: limb bones in adult ungulates. Am. Natural 109(969):547–63.

Medawar P. B., Medawar J. S. 1983. Aristotle to Zoos. A Philosophical Dictionary of Zoology. Cambridge, Mass.: Harvard Univ. Press. 305 p.

Mednikov B. M. 1980. The law of homologous series in variability. Moscow: Znanie (in Russian).

Menner V. V., Makridin V. P., eds. 1988. Sovremennaja Paleontologija [Contemporary Paleonogology]. Vol. 2. Moscow: Nedra. 384 p. (in Russian).

Meyen S. V. 1973. Plant morphology in its nomothetical aspects. Bot. Rev. 39(3):205–60.

———. 1974. On the relationship between nomogenetic and tichogenetic aspects of evolution. Zhurn. Obstchej Biologii 35(3):353–64 (in Russian).

———. 1975. The problem of directionality in evolution. In: Itogi Nauki i Tekhniki. Ser. Vertebrate Zoology. Vol. 7. Moscow: VINITI AN SSSR. pp. 66–117 (in Russian).

———. 1978. The main aspects of organismal typology. Zhurn. Obstchej Biologii 39(4): 495–508 (in Russian).

———. 1990. Non-trivial biology (remarks on . . .). Zhurn. Obstchej Biologii 51(1):4–14 (in Russian).

Miyan J. A., Ewing A. W. 1985. Is the "click" mechanism of dipteran flight an artefact of CC14 anasthesia? J. Exp. Biol. 116:313–22.

———. 1988. Further observations on dipteran flight: details of the mechanism. J. Exp. Biol. 136:229–41.

Möhl B. 1988. Short-term learning during flight control in *Locusta migratoria*. J. Comp. Physiol. 163:803–12.

Mordukhaj-Boltovskoj D. D. 1936. Geometrija Radioliarij [The Geometry of Radiolaria]. In: Uchen. Zapiski Rostovskogo Universiteta. no. 8. Rostov-on-Don: Azovo-Chernomorsk. Pub. House. pp. 3–91 (in Russian).

Mordvinov Yu. E. 1992. Kinematic peculiarities of flight in some Alcid birds as related to their ecological morphology. Zool. Zhurn. 71(7):86–92 (in Russian).

Müller A. 1972. Schuppenuntersuchungen an *Parnassius simo* (Lep., Parnassiidae). Entomol. Zeitschrift 82(18):201–10.

Nachtigall W. 1964. Zur Aerodynamik des Coleopterenfluges: Wirken die Elytren als Tragflügel? Verhdl. Dtsch. Zool. Ges. Kiel 58:319–26.

———. 1965. Die aerodynamische Funktion der Schmetterlingsschuppen. Naturwiss. 52(9):216–17.

———. 1966. Die Kinematic der Schlagflügelbewegungen von Dipteren. Methodische und analytische Grundlagen zur Biophysik der Insektenflugs. Zeitschrift Vergl. Physiol. 52(2):155–211.

———. 1967. Aerodynamische Messungen am Tragflügelsystem segelnder Schmetterlinge. Z. Vergl. Physiol. 54(2):210–31.

———. 1977. Die aerodynamische Polare des *Tipula*- Flügels und eine Einrichtung zur halbautomatischen Polarenaufnahme. In: Nachtigall W., ed. The Physiology of Movement: Biomechanics. Stuttgart: G. Fischer Verl. pp. 347–52.

———. 1979. Rasche Richtungsänderungen und Torsionen schwingender Fliegenflügel und Hypothesen über zugeordnete instationäre Strömungseffekte. J. Comp. Physiol. A 133:351–55.

———. 1980. Rasche Bewegungsänderungen bei der Flügelschwingung von Fliegen und ihre mögliche Bedeutung fur instationäre Luftkrafterzeugung. Abh. Akad. Wiss. und Lit. Math.–Naturwiss. Kl. Funktionsanal. Biol. Syst., no. 6:115–29.

———. 1981a. Der Vorderflügel grosser Heuschrecken als Luftkrafterzeuger. 1. Modellmessungen zur aerodynamischen Wirkung umterschiedlicher Flügelprofile. J. Comp. Physiol. A 142(1):127–34.

———. 1981b. Über den Einfluss von geometrischen Flügeländerungen auf die aerodynamische Funktion des Vorderflügels der Wustenheuschrecke. Eine weiterfuhrende Analyse der Jensenschen Untersuchungen. Zool. Jb. Anat. 106(1):1–11.

———. 1985. *Calliphora* as a model system for analysing insect flight. In: Kerkut G. A., Gilbert L. I., eds. Comprehensive Insect Physiology, Biochemistry and Pharmacology. Vol. 5. Oxford: Pergamon. pp. 571–605.

Nachtigall W., Roth W. 1983. Correlations between stationary measurable parameters of wing movement and aerodynamic force production in the blowfly *Calliphora vicina*. J. Comp. Physiol. A 150(2):251–60.

Nalbach G. 1988. How does *Calliphora* use the gear change mechanism during flight steering? Verhdl. Dtsch. Zool. Ges. 81:352.

———. 1989. The gear change mechanism of the blowfly (*Calliphora erythrocephala*) in tethered flight. J. Comp. Physiol. A 165(3):321–31.

Nalbach G., Hengstenberg R. 1986. Die Halteren von *Calliphora* als Drehsinnesorgan. Verhdl. Dtsch. Zool. Ges. 79:229.

Natochin Yu. V. 1988a. Certain principles of the evolution of functions on cellular, organ and organismal levels (given the example of kidney and water-salt homeostasis). Zhurn. Obstchej Biologii 49(3):291–303 (in Russian).

———. 1988b. The evolution of views on the evolution of functions. In: Darvinizm: Istorija i Sovremennost [Darwinism: History and Nowadays]. Leningrad: Nauka. pp. 130–37 (in Russian).

Navas N. 1918. Monographia de l'ordre dels Rafidiopters. Barcelona. 89 p.

Neville A. C. 1960. Aspects of flight mechanics on anisopterous dragonflies. J. Exp. Biol. 37:631–56.

Newman B. G., Savage S. B., Schouella D. 1977. Model tests on a wing section of an *Aeshna* dragonfly. In: Pedley T. J., ed. Scale Effects in Animal Locomotion. New York: Academic. pp. 445–77.

Newman D. J. S., Wootton R. J. 1986. An approach to the mechanics of pleating in dragonfly wings. J. Exp. Biol. 125:361–72.

Niculescu E. V. 1978. Les écailles androconialis chez les Polyommatinae (Lycaenidae) bons criteries spécifiques et generiques. Rev. Romaine Biol. Ser. Biol. Animale 23(1):15–19.

Niehaus M. 1981. Flight and flight control by the antennae in the small tortoiseshell (*Aglais urticae* L., Lepidoptera). II. Flight mill and free flight experiments. J. Comp. Physiol. 145:257–64.

Nielsen E. S., Davis D. R. 1981. A revision of the Neotropical Incurvariidae s. str., with the description of two new genera and two new species (Lepidoptera: Incurvarioidea). Steenstrupia 7(3):25–57.

———. 1985. The first southern hemisphere prodoxid and the phylogeny of the Incurvarioidea (Lepidoptera). Syst. Entomol. 10:307–22.

Nijhout H. F. 1991. The Development and Evolution of Butterfly Wing Patterns. Washington: Smithsonian Institution Press. 297 p.

Nikolajchuk L. A., Kuzmenko A. A., Vronsky A. A. 1991. The morphological function of locomotor apparatus of fishes. In: Sovremennaja Evoliutsionnaja Morfologija [Contemporary Evolutionary Morphology]. Kiev: Naukova Doumka. pp. 154–76 (in Russian).

Nitzsch C. L. 1818. Die Familien und Gattungen der Thierinsekten als ein Prodromus der Naturgeschichte derselben. Mag. d'Entomol. 3:261–316.

Norberg R. A. 1972a. The pterostigma of insect wings as an inertial regulator of wing pitch. J. Comp. Physiol. 81(1):9–22.

———. 1972b. Evolution of flight of insects. Zool. Scripta 1:247–50.

———. 1972c. Flight characteristics of two plume moths *Alucita pentadactyla* L. and *Orneodes hexadactyla* L. (Microlepidoptera). Zool. Scripta 1(6):247–50.

———. 1975. Hovering flight of the dragonfly *Aeshna juncea* L. In: Wu T. Y., Brokaw C. J., Brennen C., eds. Swimming and Flight in Nature. New York: Plenum. pp. 763–81.

Norris K. S., Lowe C. H. 1964. An analysis of background color-matching in amphibians and reptiles. Ecology 45(3):565–79.

Novokshonov V. G. 1992. The early evolution of caddisflies (Trichoptera). Zool. Zhurn. 71(12):58–68 (in Russian).

———. 1993. The early evolution of scorpion flies (Insecta: Panorpida). Ph.D. diss., Paleontological Institute of Russian Academy of Sciences, Moscow (in Russian).

———. 1994. The scorpion flies Permochoristidae are the nearest common ancestors of recent scorpion flies (Insecta: Panorpida = Mecoptera). Zool. Zhurn. 73(7/8):58–70 (in Russian).

Ohkubo N. 1973. Experimental studies on the tethered flight of planthoppers by the tethered flight technique. 1. Characteristics of flight of the brown planthopper, *Nilaparvata lugens* (Stal) and effects of some physical factors. Jap. J. Appl. Entomol. Zool. 17:10–18.

Ohmi K., Coutanceau M., Daube O., Loc T. P. 1991. Further experiments on vortex formation around an oscillating and translating airfoil at large incidences. J. Fluid Mech. 225:607–30.

Ohmi K., Coutanceau M., Loc T. P., Dulieu A. 1990. Vortex formation around an oscillating and translating airfoil at large incidences. J. Fluid Mech. 211:37–60.

Olivier G. A. 1789. Encyclopédie méthodique. Dictionnaire des insectes. Paris: Pankouke. Vol. 4. 331 p.; vol. 5. 793 p.

Ono T. 1977. The scales as a releaser of the copulation attempt in Lepidoptera. Naturwiss. 64 (7):386–87.

———. 1980. Role of the scales as a releaser of the copulation attempt in the silkworm moth, *Bombyx mori* (Lepidoptera, Bombycidae). Kontyu 48 (4):540–44.

Onslow H. 1921. On a periodic structure in many insect scales, and the cause of their iridescent colours. Phil. Trans. Roy. Soc. London Ser. B. 211 (382):1–74.

Oster G. F., Shubin N., Murray J. D., Alberch P. 1988. Evolution and morphogenetic rules: the shape of the vertebrate limb in ontogeny and phylogeny. Evolution 42 (5):862–84.

Packard A. S. 1886. A new arrangement of the orders of insects. Am. Natural. 20:808.

Pennycuick C. J. 1975. Mechanics of flight. In: Farner D. S., King J. R., eds. Avian Biology. Vol. 5. London: Academic P. pp. 1–75.

Pennycuick C. J., Fuller M. R., McAllister L. 1989. Climbing performance of Harris' hawks (*Parabuteo unicinctus*) with added load—implications for muscle mechanics and for radiotracking. J. Exp. Biol. 142:17–29.

Pennycuick C. J., Klaassen M., Kvist A., Lindstrom A. 1996. Wingbeat frequency and the body drag anomaly: wind-tunnel observations on a thrush nightingale (*Luscinia luscinia*) and a teal (*Anas crecca*). J. Exp. Biol. 199:2757–65.

Petukhov S. V. 1981. Biomekhanika, Bionika and Simmetrija [Biomechanics, Bionics and Symmetry]. Moscow: Nauka. 240 p. (in Russian).

Pfau H. K. 1973. Fliegt unsere Schmeissfliege mit Gangschaltung? Naturwissenschaften 60:160–61.

——. 1977. Zur Morphologie und Function des Vorderflügels und Vorderflügelgelenks von *Locusta migratoria* L. Fortschr. Zool. 24 (2/3):341–45.

——. 1978. Funktionsanatomische aspekte des Insektenflugs. Zool. Jahrb. Abt. Anat. Ontog. Tiere 99 (1):99–108.

——. 1982. Die Drehbewegungen des Libellenflügels um die Langsachse—funktionelle Anatomie und Mechanorezeption. Verh. Dtsch. Zool. Ges. p. 284.

——. 1985. Zur functionellen und phylogenetischen Bedeutung der "Gangschaltung" der Fliegen. Verh. Dtsch. Zool. Ges. 78:168.

——. 1986. Untersuchungen zur Konstruktion, Funktion und Evolution des Flugapparates der Libellen (Insecta, Odonata). Tijdschr. Entomol. 129 (3):35–123.

——. 1987. Critical comments on a "novel mechanical model of dipteran flight" (Miyan & Ewing, 1985). J. Exp. Biol. 128:463–68.

Pfau H. K., Nachtigall W. 1981. Der Vorderflügel grosser Heuschrecken als Luftkrafterzeuger. 2. Zusammenspiel von Muskeln und Gelenkmechanik bei der Einstellung der Flügelgeometrie. J. Comp. Physiol. A 142 (1):135–40.

Pianka E. 1978. Evolutionary Ecology. 2nd ed. New York: Harper & Row.

Plate L. 1928. Teorija Evoliutsii [Evolutionary Theory]. Moscow and Leningrad: GIZ. 223 p. (in Russian).

Polhamus E. C. 1971. Predictions of vortex lift characteristics by a leading-edge suction analogy. J. Aircraft 8:193–98.

Pond C. M. 1973. Initiation of flight and preflight behaviour of anisopterous dragonflies *Aeschna* spp. J. Insect Physiol. 19:2225–29.

Ponomarenko A. G. 1972. On the nomenclature of wing venation in beetles. Entomol. Obozr. 51 (4):768–75 (in Russian, published in English in Entomol. Rev. [Wash.] 1972; 51 (4):454–58).

Popov Yu. A. 1971. The historical development of hemipterans Nepomorpha infraorder (Heteroptera). Moscow: Nauka. (Trudy PIN AN SSSR. Vol. 129). 231 p. (in Russian).

Pouchkova L. V. 1971. Functioning of wings of Hemiptera and the pathways of their specialization. Entomol. Obozr. 50 (3):537–49 (in Russian, published in English in Entomol. Rev. [Wash.] 1971; 50 (3):303–9).

Prandtl L. 1952. Fluid Dynamics. New York: Hafner.

Prandtl L., Tietjens O. G. 1957. Applied Hydro- and Aeromechanics. New York: Dover. 311 p.

Pringle J. 1957. Insect Flight. Cambridge: Cambridge Univ. Press. 132 p.

Ragge D. R. 1955. The wing-venation of the order Phasmida. Trans. Roy. Entomol. Soc. London 106 (9):375–92.

Rainey R. C. 1965. The origin of insect flight: some implications of recent findings from palaeoclimatology and locust migration. In: Proc. 12th Int. Congr. Entomol., London. 8–16 July 1964. London. p. 134.

Rashevsky N. 1943a. Outline of a new mathematical approach to general biology. I. Bull. Math. Biophys. 5:33–47.

———. 1943b. Outline of a new mathematical approach to general biology. II. Bull. Math. Biophys. 5:49–73.

———. 1954. Topology and life: search of general mathematical principles in biology and sociology. Bull. Math. Biophys. 16 (4):317–48.

———. 1965. Models and mathematical principles in biology. In: Waterman T. H., Morowitz H. J., eds. Theoretical and Mathematical Biology. New York: Blaisdell.

Rasnitsyn A. P. 1966. Development of regulation ability as the cause of evolutionary progress. Bull. MOIP Div. Biol., no. 3:149–50 (in Russian).

———. 1969. Proiskhozhdenie I Evoliutsija Nizshikh Pereponchatokrylykh [The Origin and Evolution of Primitive Hymenopterans]. Moscow: Nauka 196 p. (in Russian).

———. 1971. On the problem of reasons for morpho-functional progress. Zhurn. Obstchej Biologii 32 (5):549–56 (in Russian).

———. 1972. On the problem of reasons for morpho-functional progress. In: Zakonomernosti Progressivnoj Evoliutsii [The Regularities of Progressive Evolution]. Leningrad: Nauka. pp. 314–19 (in Russian).

———. 1976a. On the early evolution of insects and the origin of Pterygota. Zhurn. Obstchej Biologii 37 (4):543–55 (in Russian with English summary).

———. 1976b. Grylloblattidae are the living members of the order Protoblattodea. Dokl. Akad. Nauk SSSR. 228:502–4 (in Russian, published in English in Doklady Biol. Sci. 228:273–75).

———. 1980. Origin and evolution of Hymenoptera. In: Trans. Paleontological Inst., Acad. Sci. USSR. Vol. 174. Moscow: Nauka Press. 192 p. (in Russian).

———. 1981. A modified paranotal theory of insect wing origin. J. Morphol. 168:331–38.

———. 1982. Proposal to regulate the names of taxa above the family group. Bull. Zoolog. Nomenclature 39:200–7.

———. 1984. Living being as adaptive trade-off. In: Mikroevoliutsia [Microevolution]. Moscow: Nauka. pp. 233–34 (in Russian).

———. 1986. Inadaptation and true adaptation. Paleontol. Zhurn., no. 1:3–7 (in Russian).

———. 1987. The rate of evolution and evolutionary theory (the hypothesis of adaptive trade-off). In: Evoliutsia i Biotsenoticheskie Krizisy [Evolution and Biocoenotic Crises]. Moscow: Nauka. pp. 46–63 (in Russian).

————. 1992a. The principles of phylogenetics and taxonomy. Zhurn. Obstchej Biologii 53 (2):176–85 (in Russian).

————. 1992b. *Strashila incredibilis*, a new enigmatic mecopteroid insect with possible siphonateran affinities from the Upper Jurassic of Siberia. Psyche 99:319–29.

————. 1996. Conceptual issues in phylogeny, taxonomy, and nomenclature. Contributions Zool. (Netherlands). 66 (1): 3–41.

————. 1997. Problem of the basal dichotomy of the winged insects. In: Fortey R. A., Thomas R. H., eds. Arthropod Relationships. Chapman & Hall. pp. 237–48.

————. On the taxonomic position of the insect order Zorotypida = Zoraptera. Zoologischer Anzeiger. Forthcoming

Rasnitsyn A. P., Dlussky G. M. 1988. The principles and methods of phylogeny reconstruction. In: Melovoj Biotsenoticheskij Krizis i Evoliutsia Nasekomykh [The Cretaceous Biocoenotic Crisis and the Evolution of Insects]. Moscow: Nauka. pp. 5–15 (in Russian).

Raup D. M. 1966. Geometric analysis of shell coiling: general problems. J. Paleontol. 40:1178–90.

————. 1967. Geometric analysis of shell coiling: coiling in ammonoids. J. Paleontol. 41:43–65.

Raup D. M., Michelson A. 1965. Theoretical morphology of the coiled shell. Science 147 (3663):1294–95.

Raup D. M., Stanley S. M. 1971. Principles of Paleontology. San Francisco: WH Freeman.

Rautian A. S. 1988. Paleontology as a source of data on regularities and factors of evolution. In: Sovremennaja Paleontologija [Contemporary Paleontology]. Vol. 2. Moscow: Nedra. pp. 76–118 (in Russian).

Rayner J. M. V. 1979a. A vortex theory of animal flight. Part 1. The vortex wake of hovering animal. J. Fluid Mech. 91:697–730.

————. 1979b. A new approach to animal flight mechanics. J. Exp. Biol. 80:17–54.

————. 1980. Vorticity and animal flight. In: Aspects of Animal Movement. Soc. Exp. Biol. Seminar Series. Vol. 5. Cambridge: Cambridge Univ. Press.

————. 1986. Vertebrate flapping flight mechanics and aerodynamics, and the evolution of flight in bats. In: Nachtigall W., ed. BIONA Rept. Mainz. Vol. 5. pp. 27–74.

————. 1991. Wake structure and force generation in avian flapping flight. In: Acta 20th Congr. Int. Ornithol. Wellington. pp. 702–15.

Rayner J. M. V., Aldridge H. D. J. N. 1985. Three-dimensional reconstruction of animal flight paths and the turning flight of microchiropteran bats. J. Exp. Biol. 118:247–66.

Rayner J. M. V., Jones G., Thomas A. 1986. Vortex flow visualizations reveal change in upstroke function with flight speed in bats. Nature 321 (6066):162–64.

Rees C. J. 1975. Aerodynamic properties of an insect wing section and a smooth aerofoil compared. Nature 258 (5531):141–42.

Reeve E. C. R., Huxley J. S. 1945. Some problems in the study of allometric growth. In: Essays on Growth and Form. Oxford Univ. Press. pp. 121–56.

Remington C. L. 1955. The Apterygota. In: A Century Progress in Natural Sciences. Centennial vol. California Academy of Sciences. pp. 795–505.

Riedl R. 1977. A systems-analytical approach to macroevolutionary phenomena. Quart. Rev. Biol. 52 (4):351–70.

Riek E. F., Kukalova-Peck J. 1984. A new interpretation of dragonfly wing venation based upon Early Upper Carboniferous fossils from Argentina (Insecta: Odonatoidea) and basic character states in pterygote wings. Can. J. Zool. 62:1150–66.

Robertson R. M., Johnson A. G. 1993. Collision avoidance of flying locusts: steering torques and behaviour. J. Exp. Biol. 183:35–60.

Robertson R. M., Reye D. N. 1992. Wing movements associated with collision-avoidance manoeuvres during flight in the locust *Locusta migratoria*. J. Exp. Biol. 163:231–58.

Robinson M. C., Luttges M. W. 1983. Unsteady flow separation and attachment induced by pitching airfoils. American Institute of Aeronautics and Astronautics, paper 83-0131. 14 p.

Rohdendorf B. B. 1949. Evoliutsia i Klassifikatsia Letatel'nogo Apparata Nasekomykh [The Evolution and Classification of Insect Flight Systems]. Moscow and Leningrad: Izd. AN SSSR. pp. 1–176 (in Russian).

———. 1961. Nadotriad Psocopteroidea. Senoedoobraznye [Superorder Psocopteroidea]. In: Rohdendorf B. B., ed. Osnovy paleontologii. Chlenistonogie. Trakheynye i Khelitserovye. Moscow: Nauka 226 p. (in Russian); English translation (with a wrong authorship!) In: Rohdendorf B. B., ed. 1991. Fundamentals of Paleontology. Vol. 9. Arthropoda, Tracheata, Chelicerata. Washington, D.C.: Smithsonian Institution Libraries and National Scientific Foundation. 317 p.

———. 1968. Napravlenija Filogeneticheskogo Razvitija Krylatyh Nasekomykh [Trends in Pterygote Phylogeny]. Zhurn. Obstchej Biologii 29:57–67 (in Russian).

———. 1977. The rationalization of names of higher taxa in zoology. Paleontol. Zhurn. No. 3:11–21 (in Russian, published in English in Paleontol. J., no. 11:149–55).

Rohdendorf B. B., Rasnitsyn A. P., eds. 1980. Istoricheskoe Razvitie Klassa Nasekomykh [Historical Development of the Class of Insects]. Moscow: Nauka. 270 p. (in Russian).

Rosen R. 1967. Optimality Principles in Biology. London: Butterworth.

Ross H. H. 1965. A Textbook of Entomology. 3rd ed. New York: Wiley. 539 p.

Ross H. H., Ross C. A., Ross J. R. P. 1982. A Textbook of Entomology. New York: Wiley.

Rossow V. J. 1978. Lift enhancement by an externally trapped vortex. J. Aircraft 115:618–25.

Roth G., Nishikawa K. C., Naujoks-Manteuffel C., Schmidt A., Wake D. B. 1993. Pedomorphosis and simplification in the nervous system of salamanders. Brain Behav. Evol. 42:137–70.

Rudolph R. 1976a. Preflight behaviour and the initiation of flight in tethered and unrestrained dragonfly, *Calopteryx splendens* (Harris) (Zygoptera: Calopterygidae). Odonatologica 5 (1):59–64.

———. 1976b. Die aerodynamischen Eigenschaften von *Calopteryx splendens* (Harris) (Zygoptera: Calopterygidae). Odonatologica 150 (4):383–86.

Rüppell G. 1985. Kinematic and behavioural aspects of flight of the male banded agrion, *Calopteryx (Agrion) splendens* L. In: Insect Locomotion, Proc. Symp. 17th Int. Congr. Entomol. Berlin-Hamburg. pp. 195–204.

———. 1989. Kinematic analysis of symmetrical flight manoevres of Odonata. J. Exp. Biol. 144:13–42.

Russell E. S. 1916. Form and Function. A Contribution to the History of Animal Morphology. London: J. Murray. 383 p.

Ryazanova G. I. 1966. Comparative characteristics of flight of dragonflies. Zhurn. Obstchej Biologii 27 (3):349–59 (in Russian).

———. 1968. Functional assessment of wing shape in dragonflies. In: Proc. 13th Int. Entomol. Congress, Moscow. 2–9 Aug. 1968. Vol. 1, pp. 293–94 (in Russian).

Saffman P. G. 1992. Vortex Dynamics. Cambridge, U.K.: Cambridge Univ. Press. 311 p.

Sahal D. 1976. Homeorhetic regulation and structural stability. Cybernetica 19 (4):305–15.

Saharon D., Luttges M. 1987. Three-dimensional flow produced by a pitching-plunging model dragonfly wing. American Institute of Aeronautics and Astronautics, paper 87-0121. 17 p.

———. 1988. Visualization of unsteady separated flow produced by mechanically driven dragonfly wing kinematics model. American Institute of Aeronautics and Astronautics, paper 88-0569. 23 p.

Sanderson M. J., Donoghue M. J. 1989. Patterns of variation in levels of homoplasy. Evolution 43 (8):1781–95.

Sato M., Azuma A. 1997. The flight performance of a damselfly *Ceriagrion melanurum* Selys. J. Exp. Biol. 200 (12):1765–79.

Savage S. B., Newman B. G., Wong D. T.-M. 1979. The role of vortices and unsteady effects during the hovering flight of dragonflies. J. Exp. Biol. 83:59–77.

Savchenko Yu. N. 1971. Some hydrodynamic peculiarities of propelling agent of the flapping wing type. In: Bionika. Kiev: Naukova Doumka. No. 5, pp. 11–19 (in Russian).

Saveljev S. V. 1993. Monsters. Priroda 10:55–65 (in Russian).

Saxen L., Toivonen S. 1962. Primary Embryonic Induction. London: Logos Press, 271 p.

Schlichting H. 1974. Teorija Pogranichnogo Sloja [The Theory of Boundary Layer]. Moscow: Nauka. 712 p. (in Russian).

Schmalhausen I. I. 1938. Osnovy Sravnitelnoj Anatomii Pozvonochnykh Zhivotnykh [The Foundations of Vertebrate Animals Comparative Anatomy]. Moscow: Gosudarstvennoe Uchebno-Pedagogicheskoe Izdatel'stvo Narkomprosa RSFSR. 488 p. (in Russian).

———. 1945. The problem of stability of organic form in the process of evolution. Zhurn. Obstchej Biologii 6 (1):3–25 (in Russian).

———. 1949. Factors of Evolution. Toronto: Blakiston. 327 p.

———. 1969. Problemy Darvinizma [Problems of Darwinism]. Leningrad: Nauka. 493 p. (in Russian).

———. 1983. Puti i Zakonomernosti Evoliutsionnogo Protsessa [The Ways and Regularities of Evolutionary Process]. Moscow: Nauka. 360 p. (in Russian).

Schmidt J., Zarnack W. 1987. The motor pattern of locusts during visually induced rolling in long-term flight. Biol. Cybern. V. 56:397–410.

Schmidt K., Paulus H. 1970. Die Feinstruktur der Flügelschuppen einiger Lycaeniden (Insecta: Lepidoptera). Z. Morphol. Tiere 66 (3):224–41.

Schmidt-Nielsen K. 1972. Locomotion: energy cost of swimming, flying, and running. Science 177 (4045):222–28.

———. 1979. Animal Physiology. 2nd ed. Cambridge: Cambridge Univ. Press.

———. 1984. Scaling. Why Is Animal Size So Important? Cambridge: Cambridge Univ. Press.

Schneider P. 1975. Die Flugtypen der Käfer (Coleoptera). Entomol. German 1:222–31.

———. 1978. Die Flug- und Faltingstypen der Käfer (Coleoptera). Zool. Jb. Anat. 99:174–210.

Schneider P., Hermes M. 1976. Die Bedeutung der Elytren bei Vertretern des Melolontha-Flugtyps (Coleoptera). J. Comp. Physiol. A 106 (1):39–49.

Schneider P., Krämer B. 1974. Die Steuerung des Fluges beim Sandlaufkäfer (Cicindela) und beim Mailkäfer (Melolontha). J. Comp. Physiol. A 91 (4):377–86.

Schnell G. D. 1974. Flight speeds and wingbeat frequencies of the magnificent frigatebird. Auk 91 (July):564–70.

Schreck S. J., Luttges M. W. 1988. Unsteady separated flow structure: extended K range and oscillations through zero pitch angle. American Institute of Aeronautics and Astronautics, paper 88-0325. 14 p.

Schwenne T. 1990. Kinematische Analyse des Flügelschlags bei Rollmanövern von Wüstenheuschrecken (Schistocerca gregaria). Dissertation zur Erlangung des Doktorgrades. Gottingen. 99 p.

Schwenne T., Zarnack W. 1987. Movements of the hindwings of Locusta migratoria, measured with miniature coils. J. Comp. Physiol. A 160 (5):657–66.

Scriven M. 1959. Explanation and prediction in evolutionary theory. Science 130 (3374): 477–82.

Seilacher A. 1974. Fabricational noise in adaptive morphology. Syst. Zool. 22:451–65.

Sellier R. 1971. Étude morphologique en microscopie electronique à balayage de quelques types d'androconies alaires chez les lépidoptères diurnes. C. R. Acad. Sci. D 273 (25):2550–53.

———. 1972. Étude ultrastructurale en microscopie electronique a balayage et essai d'interpretation du mode de fonctionnement des poils androconiaux alaires chez les Hesperiidae (Lepidoptera: Rhopalocera). C. R. Acad. Sci. D 275 (20):2239–42.

———. 1973a. Contribution a l'étude de l'ultrastructure et du mode de fonctionnement de l'appareil androconial alaire chez les satyrides (lépidoptères rhopaloceres). Alexanor 8 (2):65–70.

———. 1973b. Recherches en microscopie l'electronique par balayage, sur l'ultrastructure de l'appareil androconial alaire dans le genre Argynnis et dans les genres voisins (lep. rhopaloceres nymphalides). Ann. Soc. Entomol. France 9 (3):703–28.

Send W. 1992. The mean power of forces and moments in unsteady aerodynamics. Zeitschrift für angewandte Mathematik und Mechanik. 72:113–32.

Serebrovsky A. S. 1973. Nekotorye Problemy Organicheskoj Evoliutsii [Some Problems of Organic Evolution]. Moscow: Nauka. 168 p.

Severtzov A. N. 1939. Morphological regularities of evolution. Moscow-Leningrad: Izd-vo AN SSSR. pp. 53–610 (in Russian).

Severtzov A. S. 1987. Osnovy Teorii Evoliutsii [Baselines of the Theory of Evolution]. Moscow: Izd-vo MGU. 320 p. (in Russian).

———. 1990. Napravlennost Evoliutsii [The Directionality of Evolution]. Moscow: Izd-vo MGU. 272 p. (in Russian).

Sharov A. G. 1957. Peculiar Paleozoic wingless insects of the new order Monura (Insecta, Apterygota). Dokl. Akad. Nauk SSSR 115:795–98 (in Russian).

———. 1966. On the position of the orders Glosselytrodea and Caloneurodea in the system of Insecta. Paleontol. Zhurn., no. 3:84–93 (in Russian).

———. 1968. Filogenija Ortopteroidnykh Nasekomykh [Phylogeny of Orthopteroid Insects]. Moscow: Nauka (Trudy PIN AN SSSR. Vol. 135). 218 p. (in Russian).

Sharp D. 1899. Some points in the classification of the Insecta, Hexapoda. In: Proc. 4th Int. Congr. Zool. Cambridge. pp. 246–49.

Sharplin J. 1963a. Wing base structure in Lepidoptera. 1. Fore wing base. Can. Entomol. 95 (10):1024–50.

———. 1963b. Wing base structure in Lepidoptera. 2. Hind wing base. Can. Entomol. 95 (11):1121–45.

Shcherbakov D. E., Lukashevich E. D., Blagoderov V. A. 1995. Triassic Diptera and initial radiation of the order. Int. J. Dipterologic. Res. 6 (2):75–115.

Shear W. A., Bonamo P. M., Griedson J. D., Rolf W. D. I., Smith E. L., Norton R. A. 1984. Early land animals in North America: evidence from Devonian age arthropods from Gilboa, New York. Science 224:492–94.

Shimansky V. N. 1987. Historical development of biosphere. In: Evoliutsia i Biotsenoticheskie Krizisy [Evolution and Biocoenotic Crises]. Moscow: Nauka. pp. 5–45 (in Russian).

Shimoyama I., Fujisawa Y. K., Getzan G. D., Miura H., Shimada M., Matsumoto Y. 1995. Fluid dynamics of microwing. In: Proc. IEEE Workshop on MEMS'95. pp. 380–85.

Shipley A. E. 1904. The orders of insects. Zool. Anzeiger 27:259–62.

Shishkin M. A. 1987. Individual development and the evolutionary theory. In: Evoliutsia i Biotsenoticheskije Krizisy [Evolution and Biocoenotic Crises]. Moscow: Nauka. pp. 76–123 (in Russian).

———. 1988. Evolution as an epigenetic process. In: Sovremennaja Paleontologija [Contemporary Paleontology]. Vol. 2. Moscow: Nedra. pp. 142–69 (in Russian).

Shreider Yu. A. 1983. Taxonomy, typology, classification. In: Teoria i Metodologia Biologicheskikh Klassifikatsij [Theory and Methodology of Biological Classification]. Moscow: Nauka. pp. 90–100 (in Russian).

Shtackelberg A. A. 1969. The family key. In: Opredelitel Nasekomykh Evropejskoj Chasti SSSR [The Key of Insects of the European Part of the USSR]. Vol. 5. Leningrad: Nauka. Diptera, Siphonaptera; ch. 1. Fam. Trichoceridae-Phoridae. pp. 35–55 (in Russian).

Shvets A. I., Zakharenkov M. N., Osminin P. K. 1979. Otchet Instituta Mekhaniki MGU

[Report of the Moscow University Institute of Mechanics]. Moscow: MGU (in Russian).

Shwanvich B. N. 1946. On the relationships between the orders of insects as related to the origin of flight. 1. Division of Pterygota into complexes of orders. Zool. Zhurn. 25 (6):529–42 (in Russian).

———. 1948. On the relationships between the orders of insects as related to the origin of flight. 2. Division of Pterygota into groups of orders according to the height of their organization. Zool. Zhurn. 27 (2):137–48 (in Russian).

———. 1949. Kours Obstchej Entomologii [A Course of General Entomology]. Moscow and Leningrad: Sov. Nauka. 900 p. (in Russian).

Shwartz S. S. 1968. The principle of optimal phenotype (towards the theory of stabilizing selection). Zhurn. Obstchej Biologii 29 (1):12–24 (in Russian).

———. 1980. Ekologicheskije Zakonomernosti Evoliutsii [Ecological Regularities of Evolution]. Moscow: Nauka. 278 p. (in Russian). (See also a monograph in English: Shvarts S. S. 1977. The Evolutionary Ecology of Animals. New York: Consultants Bureau. 292 p.)

Shwartz S. S., Istchenko V. G., Dobrinskaja L. A., Amstislavsky A. Z., Brousynina I. N., Paraketsov I. A., Yakovleva A. S. 1968a. The growth speed and size of fish brain (towards the problem "Species and intraspecific categories in different classes of vertebrates"). Zool. Zhurn. 47 (6):901–15 (in Russian).

Shwartz S. S., Smirnov V. S., Dobrinsky L. N. 1968b. Metod Morfo-Fiziologicheskikh Indikatorov [The Method of Morfo-Physiological Indices]. Sverdlovsk: Institut Ekologii Rastenij i Zhivotnykh (in Russian).

Silvestri F. 1913. Descrizione di un nuovo ordine di Insetti. Boll. Lab. Zool. Gen. e Agrar. della R. scuola superiore di agricultura in Portici. 7:193–209.

Simpson G. G. 1944. Tempo and Mode in Evolution. New York: Columbia Univ. Press.

———. 1980. Splendid Isolation. New Haven, Conn.: Yale Univ. Press.

Snodgrass R. E. 1935. Principles of Insect Morphology. New York: McGraw-Hill. 667 p.

Somps C., Luttges M. 1985. Dragonfly flight: novel uses of unsteady separated flows. Science 228:1326–29.

Sotavalta O. 1947. The flight-tone (wing-stroke frequency) of insects. Acta Entomol. Fenn. (Helsinki) no. 4:1–117.

———. 1952. The essential factor regulating the wingstroke frequency of insects in wing mutilation and leading experiments and in experiments at subatmospheric pressure. Ann. Zool. Soc. "Vanamo" 15:1–67.

———. 1954. The effect of wing inertia on the wing-stroke frequency of moths, dragonflies and cockroach. Ann. Entomol. Fenn. 20:93–101.

Soule M. E., ed. 1987. Viable Populations for Conservation. Cambridge: Cambridge Univ. Press.

Spedding G. R. 1986. The wake of a jackdaw (Corvus monedula) in slow flight. J. Exp. Biol. 125:287–307.

———. 1987. The wake of a kestrel (Falco tinnunculus) in flapping flight. J. Exp. Biol. 127:59–78.

————. 1992. The aerodynamics of animal flight. In: Advances in Comparative and Environmental Physiology. Vol. 11. London: Springer. pp. 51–111.

————. 1993. On the significance of unsteady effects in the aerodynamic performance of flying animals. Contemp. Mathemat. 141:401–19.

Spedding G. R., Maxworthy T. 1986. The generation of circulation and lift in a rigid two-dimensional fling. J. Fluid Mech. 165:247–72.

Spedding G. R., Rayner J. M. V., Pennycuick C. J. 1984. Momentum and energy in the wake of a pigeon (*Columba livia*) in slow flight. J. Exp. Biol. 111:81–102.

Spüler M., Heide G. 1978. Simultaneous recordings of torque, thrust and muscle spikes from the fly *Musca domestica* during optomotor responses. Zeitschrift für Naturforschung 33:455–57.

Srygley R. B., Dudley R. 1993. Correlations of the position of center of body mass with butterfly escape tactics. J. Exp. Biol. 174:155–66.

Stary J. 1990. Anagenesis of the radial field of wing in Tipulomorpha. In: 2nd Int. Congr. Dipterol., Bratislava. 27 Aug.–1 Sept. 1990. Abstr. vol. Bratislava. pp. 227.

Stephens J. F. 1829. A Systematic Catalogue of British Insects. Parts 1 and 2. London: Baldwin & Craddock. 804 p.

Stevens P. S. 1973. Space, architecture, and biology. Syst. Zool. 22 (4):405–8.

Sukacheva I. D. 1976. Caddisflies of the Permotrichoptera order. Paleontol. Zhurn., no. 2:94–105 (in Russian).

————. 1982. Historical development of the caddisflies order. Moscow: Nauka. 112 p.

Sunada S., Kawachi K., Watanabe I., Azuma A. 1993a. Fundamental analysis of three-dimensional 'near' fling. J. Exp. Biol. 183:217–48.

————. 1993b. Performance of a butterfly in take-off flight. J. Exp. Biol. 183:249–78.

Svidersky V. L. 1962. O Dialektike Elementov i Struktury v Objektivnom Mire i Poznanii [On the Dialectics of Elements and Structures in Virtual World and Cognition]. Moscow: Nauka. 275 p. (in Russian).

————. 1973. Nejrofiziologija Poliota Nasekomykh [Neurophysiology of Insect Flight]. Leningrad: Nauka. 216 p. (in Russian).

Swartz S., Biewener A. A. 1992. Shape and scaling. In: Biewener A. A., ed. Biomechanics. Structures and Systems. Oxford Univ. Press. pp. 21–43.

Taylor C. P. 1981. Contribution of compound eyes and ocelli to steering of locusts in flight. 1. Behavioural analysis. J. Exp. Biol. 93:1–18.

Taylor C. R., Weibel E. R. 1981. Design of the mammalian respiratory system. 1. Problem and strategy. Respir. Physiol. 44:1–10.

Taylor C. R., Weibel E. R., Weber J.-M., Vock R., Hoppeler H., Roberts T. J., Brichon G. 1996. Design of the oxygen and substrate pathways. 1. Model and strategy to test symmorphosis in a network structure. J. Exp. Biol. 199 (8):1643–49.

Thomas R. D. K. 1978. Limits to opportunism in the evolution of the Arcoida (Bivalvia). Phil. Trans. R. Soc. London B 284:335–44.

————. 1979. Morphology, constructional. In: Fairbridge R. W., Jablonski D., eds. Encyclopedia of Earth Sciences, vol. 7: The Encyclopedia of Paleontology. Stroudsburg, Pa.: Dowden, Hutchinson & Ross. pp. 482–87.

Thomas R. D. K., Reif W.-E. 1993. The skeleton space: a finite set of organic designs. Evolution 47 (2):341–60.

Thomson A. J., Thompson W. A. 1977. Dynamics of a bistable system: the click mechanism in Dipteran flight. Acta Biotheor. 26 (1):19–29.

Thüring D. A. 1986. Variability of motor output during flight steering in locusts. J. Comp. Physiol. A 158:653–64.

Tikhomirova A. L. 1991. Perestroika Ontogeneza kak Mekhanizm Evoliutsii Nasekomykh [Transformation of Ontogeny as Mechanism of Insect Evolution]. Moscow: Nauka. 169 p.

Tillyard R. J. 1919a. Mesosoic insects of Queensland. No 5. Mecoptera, the new order Paratrichoptera, and additions to Planipennia. Proc. Linn. Soc. New South Wales 44:194–212.

———. 1919b. A fossil insect wing belonging to the new order Paramecoptera, ancestral to the Trichoptera and Lepidoptera, from the upper Coal-Measures of Newcastle, N.S.W. Proc. Linn. Soc. New South Wales 44:231–56.

———. 1925. A new fossil insect wing from Triassic beds near Deewhy, N.S.W. Proc. Linnean Soc. New South Wales 50:374–77.

———. 1928. Kansas Permian insects. Part 10. The new order Protoperlaria: a study of the typical genus *Lemmatophora* Sellards. Am. J. Sci. (ser. 5). 16:185–220.

———. 1931. Kansas Permian insects. Part 13. The new order Protelytroptera, with a discussion of its relationships. Am. J. Sci. (ser. 5). 21:232–66.

Timofeeff-Ressovsky N. V. 1995. Vospominanija [Reminiscences]. Moscow: Pangeja. 383 p.

Timofeeff-Ressovsky N. V., Vorontzov N. N., Yablokov A. V. 1977. Kratkij Ocherk Teorii Evoliutsii [Brief Essay of the Theory of Evolution]. Moscow: Nauka. 302 p. (in Russian).

Tobalske B. W., Dial K. P. 1996. Flight kinematics of black-billed magpies and pigeons over a wide range of speeds. J. Exp. Biol. 199 (2):263–80.

Tobias V. I. 1992. Links of some taxonomical characters (wing venation, coloration, etc.) of Hymenoptera with habitat. In: 19th Int. Congr. Entomol., Beijing. 28 June–4 July 1992. Proc. Abstr. Beijing. pp. 215.

———. 1993. The dependence of wing venation of hymenopteran insects (Hymenoptera) on their environment. Entomol. Obozr. 72 (3):497–506 (in Russian).

Trueman J. W. H. 1990. Comment—Evolution of insect wings: a limb exite plus endite model. Can. J. Zool. 68 (6):1333–35.

Tsvelykh A. N. 1986. The relationship between flight speed and wing stroke frequency in free flight of sandwich tern *Thalasseus sandvicensis* Lath. Dokl. AN SSSR 6 (8):82–83 (in Russian).

———. 1988. Flight speed, wing stroke frequency and the energetics of flying slender-billed gulls. Vestn. Zool. No. 3:41–45 (in Russian).

Tucker V. A. 1966. Oxygen consumption of a flying bird. Science 154:150–51.

Turner W. J., Babcock J. M., Jenkins J. 1986. New record and first observations of adult flight activity for *Deuterophlebia coloradensis* Pennak (Diptera: Deuterophlebiidae) in Idaho. Pan-Pacif. Entomol. 62 (2):111–18.

Tystchenko V. P. 1992. Vvedenije v Teoriju Evoliutsii [Introduction to the Evolutionary Theory]. SPb: Izd-vo SPbGU. 240 p. (in Russian).

Ugolev A. M. 1985. Evoliutsia Pistchevarenija i Printsipy Evoliutsii Funktsij. Elementy Sovremennogo Funktsionalizma [The Evolution of Digestion and Principles of Evolution of Functions. The Elements of Modern Functionalism]. Leningrad: Nauka. 544 p. (in Russian).

Uldrick J. P. 1968. On the propulsion efficiency of swimming flexible hydrofoils of finite thickness. J. Fluid Mech. 32:29–53.

Unwin D. M., Corbet S. A. 1984. Wingbeat frequency, temperature and body size in bees and flies. Physiol. Entomol. 9 (1):115–21.

Ushatinskaya R. S. 1987. Thermoregulation in the class of insects. In: Voprosy Ekologcheskoj Fiziologii Nasekomykh [Problems of Insect Ecological Physiology]. Moscow: Nauka. pp. 5-46 (in Russian).

Ussatchov D. A. 1970. Two main factors in the formation of the wing venation of Diptera. Beitr. Entomol. 20 (1–2):5–10.

Vakhrushev A. A., Rautian A. S. 1993. Historical approach to the ecology of communities. Zhurn. Obstchej Biologii 54 (5):532–53 (in Russian).

van den Berg C., Ellington C. P. 1997a. The three-dimensional leading-edge vortex of a 'hovering' model hawkmoth. Phil. Trans. Roy. Soc. London Ser. B. 352:329–340.

———. 1997b. The vortex wake of a 'hovering' model hawkmoth. Phil. Trans. Roy. Soc. London Ser. B. 352:317–328.

Van Valen L. M. 1974. A natural model for the origin of some higher taxa. J. Herpetol. 8 (2):109–21.

Vasiutinsky N. A. 1990. Zolotaja Proportsia [The Golden Proportion]. Moscow: Molodaja Gvardija. 240 p. (in Russian).

Vavilov N. I. 1968. The law of homologous series in hereditary variability. In: Klassiki Sovetskoj Genetiki [The Classics of Soviet Genetics]. Leningrad: Nauka. pp. 9–50 (in Russian, published in English in J. Genet. 1922; 12 (1):47–90).

Vedenov M. F., Kremiansky V. I., Shatalov A. T. 1972. The concept of structural levels in biology. In: Razvitie Kontseptsii Strukturnykh Urovnej v Biologii [Development of the Concept of Structural Levels in Biology]. Moscow: Nauka. pp. 7–70 (in Russian).

Verhoeff K. W. 1904. In: Nova Acta der Kaiserliche Leopold-Carol Deutschen Akademie Naturforscher (Halle). Vol. 84. p. 109.

Vermeij G. J. 1987. Evolution and Escalation. An Ecological History of Life. N.J.: Princeton Univ. Press. 528 p.

———. 1995. Economics, volcanoes, and Phanerozoic revolutions. Paleobiology 21 (2):125–52.

Videler J. J., Groenewegen A., Gnodde M., Vossebelt G. 1988. Indoor flight experiments with trained kestrels. 2. The effect of added weight on flapping flight kinematics. J. Exp. Biol. 134:185–99.

Vielmetter W. 1958. Physiologie des Verhaltens zur Sonnenstrahlung bei den Tagfalter Argynnis paphia L. 1. Untersuchungen im Freiland. J. Insect Physiol. 2 (1):13–37.

Vogel S. 1962. A possible role of the boundary layer in insect flight. Nature 193(4821):1201–2.

————. 1965. Aspects of flight at low Reynolds number. In: Proc. 12th Int. Congr. Entomol., London. 8–16 July 1964. London. pp. 188–89.

————. 1966. Flight in *Drosophila*. 1. Flight performance of tethered flies. J. Exp. Biol. 44 (3):567–78.

————. 1967a. Flight in *Drosophila*. 2. Variations in stroke parameters and wing contour. J. Exp. Biol. 46 (2):383–92.

————. 1967b. Flight in *Drosophila*. 3. Aerodynamic characteristics of fly wings and wing models. J. Exp. Biol. 46 (3):431–43.

Vogel S., Feder N. 1966. Visualization of low-speed flow using suspended plastic particles. Nature 209 (5019):186–87.

von Laicharting J. N., 1781. Verzeichnis und Beschreibung der Tyroler Insekten. Table 1. Zurich: Flüessly. 283 p.

von Stummer-Traunfels R., 1891. Vergleichende Untersuchungen über die Mundwerkzeuge der Thysanuren und Collembolen. Sitzungsberichten der Akademie Wien C (1). pp. 216–36.

Vorobjov N. N. 1991. The effect of generation of transversal force under a flow over a rough surface. Dokl. AN SSSR 318 (1):62–65 (in Russian).

Vorobjova E. I. 1980. Parallelism and convergence in the evolution of Coelacanthimorpha fishes. In: Morfologicheskie Aspekty Evoliutsii [Morphologic Aspects of Evolution]. Moscow: Nauka. pp. 7–28 (in Russian).

————. 1991. Expediency and stability of evolutionary morphological transformations. In: Sovremennaja Evoliutsionnaja Morfologija [Contemporary Evolutionary Morphology]. Kiev: Naukova Doumka. pp. 56–71 (in Russian).

Vorobjova E. I., Meyen S. V. 1988. Morphological research in paleontology. In: Sovremennaja Paleontologija [Contemporary Paleontology]. Vol. 1. Moscow: Nedra. pp. 80–122 (in Russian).

Vorontzov N. N. 1961. Nonuniformity of the rate of digestive organs transformation in rodents and the principle of compensation of functions. Dokl. AN SSSR 136 (6):1494–97 (in Russian).

————. 1967. Evoliutsia Pistchevaritelnoj Sistemy Gryzunov [The Evolution of Digestion System in Rodents]. Novosibirsk: Nauka. 239 p. (in Russian).

————. 1989. On the methodology of morphology and the levels of morphological analysis. Zhurn. Obstchej Biologii 50 (6):737–45 (in Russian).

Waddington C. 1940. Organizers and Genes. Cambridge: Cambridge Univ. Press.

Wagner H. 1982. Flow-field variables trigger landing in flies. Nature 297 (5862):147–48.

Wainwright S. A. 1988. Axis and Circumference. The Cylindrical Shape of Plants and Animals. Cambridge, Mass.: Harvard Univ. Press. 132 p.

Wake D. B. 1991. Homoplasy: The result of natural selection, or evidence of design limitations? Am. Natural. 138 (3):543–67.

————. 1996. Introduction. In: Sanderson M. J., Hufford L., eds. Homoplasy. The Recurrence of Similarity in Evolution. San Diego: Academic. pp. xvii–xxv.

Wakeling J. M. 1993. Dragonfly aerodynamics and unsteady mechanisms: a review. Odonatologica 22 (3):319–34.

Wakeling J. M., Ellington C. P. 1997a. Dragonfly flight. I. Gliding flight. J. Exp. Biol. 200:543–56.

———. 1997b. Dragonfly flight. II. Velocities, accelerations and kinematics of flapping flight. J. Exp. Biol. 200:557–82.

———. 1997c. Dragonfly flight. III. Lift and power requirements. J. Exp. Biol. 200:583–600.

Waldmann B., Zarnack W. 1987. Motor activity and movements of forewings during roll manoeuvres in flying desert locusts. In: Elsner N., Creutzfeldt O., eds. New Frontiers in Brain Research. Stuttgart: Thieme. Poster 54.

———. 1988. Forewing movements and motor activity during roll manoeuvres in flying desert locust. Biol. Cybern. 59 (4–5):325–35.

Walker E. M. 1914. A new species of Orthoptera forming a new genus and family. Can. Entomol. 46:93–99.

Waloff Z. 1972. Observations on the airspeeds of freely flying locusts. Anim. Behav. 20 (2):367–72.

Ward J. P., Baker P. S. 1982. The tethered flight performance of a laboratory colony of *Triatoma infestans* (Klug.) (Hemiptera: Reduviidae). Bull. Entomol. Res. 75 (1):17–28.

Wasserthal L. T. 1974. Function und Entwicklung der Flügel der Federmotten (Lepidoptera, Pterophoridae). Z. Morphol. Tiere 77 (2):127–55.

———. 1975. The role of butterfly wings in regulation of body temperature. J. Insect Physiol. 21 (12):1921–30.

Watt W. B. 1968. Adaptive significance of pigment polymorphisms in *Colias* butterflies. 1. Variation in melanin pigment in relation to thermoregulation. Evolution 22:437–58.

Weibel E. R., Taylor C. R., Hoppeler H. 1991. The concept of symmorphosis: A testable hypothesis of structure function relationship. Proc. Natl. Acad. Sci. USA 88 (22):10357–61.

Weibel E. R., Taylor C. R., Weber J.-M., Vock R., Roberts T. J., Hoppeler H. 1996. Design of the oxygen and substrate pathways. VII. Different structural limits for oxygen and substrate supply to muscle mitochondria. J. Exp. Biol. 199 (8):1699–1709.

Weis-Fogh T. 1956a. Biology and physics of locust flight. II. Flight performance of the desert locust (*Schistocerca gregaria*). Phil. Trans. Roy. Soc. London B 239(667):459–510.

———. 1956b. Biology and physics of locust flight. IV. Notes on sensory mechanisms in locust flight. Phil. Trans. Roy. Soc. London B 239 (667):553–84.

———. 1959. Elasticity in arthropod locomotion: a neglected subject, illustrated by the wing system of insects. In: 25th Int. Congr. Zool. Vol. 4. pp. 393–95.

———. 1960. A rubber-like protein in insect cuticle. J. Exp. Biol. 37 (4):889–907.

———. 1964. Control of basic movements in flying insects. Symp. Soc. Exp. Biol. Vol. 18. Cambridge: Cambridge Univ. Press. pp. 343–61.

———. 1965. Elasticity and wing movements in insects. In: Proc. 12th Int. Congr. Entomol., London. 8–16 July 1964. London. pp. 186–88.

———. 1973. Quick estimates of flight fitness in hovering animals, including novel mechanisms for lift production. J. Exp. Biol. 59:169–230.

————. 1975. Flapping flight and power in birds and insects, conventional and novel mechanisms. In: Wu T. Y., Brokaw C. J., Brennan C., eds. Swimming and Flight in Nature. Vol. 1. New York: Plenum. pp. 729–62.

Weis-Fogh T., Andersen S. O. 1970. New molecular model for the long-range elasticity of resilin. Nature 227:718–21.

Weis-Fogh T., Jensen M. 1956. Biology and physics of locust flight. 1. Basic principles in insect flight. A critical review. Phil. Trans. Roy. Soc. London B 239 (667):415–58.

————. 1993. Ecological correlates of hovering flight of hummingbirds. J. Exp. Biol. 178:59–70.

Went F. W. 1971. Parallel evolution. Taxon 20 (213):197–226.

White J. F., Gould S. J. 1965. Interpretation of the coefficient in the allometric equation. Am. Natural. 99 (904):5–18.

Wigglesworth V. B. 1963. Origin of wings in insects. Nature 197:97–98.

————. 1973. Evolution of insect wings and flight. Nature 246(5429):127–29.

————. 1976. The evolution of insect flight. In: Insect Flight. Rainey R. C., ed. Oxford: Blackwell Scientific. pp. 255–69.

Wilkin P. 1990. The instantaneous force on a desert locust, *Schistocerca gregaria*, flying in a wind tunnel. J. Kansas Entomol. Soc. 63 (2):316–28.

Wilkin P. J. 1985. Aerodynamics. In: Kerkut G. A., Gilbert L. I., eds. Comprehensive Insect Physiology, Biochemistry and Pharmacology. Vol. 150. Oxford: Pergamon. pp. 553–70.

Willmott A. P., Ellington C. P., Thomas A. L. R. 1997. Flow visualization and unsteady aerodynamics in the flight of the hawkmoth *Manduca sexta*. Phil. Trans. Roy. Soc. London Ser. B. 352:303–16.

Wilson D. M. 1965. Nervous control of insect flight. In: Proc. 12th Int. Congr. Entomol., London. 8–16 July 1964. London. pp. 192–93.

Wisser A. 1988. Wing beat of *Calliphora erythrocephala*: turning axis and gearbox of the wing base (Insecta, Diptera). Zoomorphology 107 (6):359–69.

Wisser A., Nachtigall W. 1983. Funktionelle Gelenkmorphologie und Flügelantrieb bei der Schmeissflige. In: Nachtigall W., ed. BIONA Rept. No. 1, pp. 29–34.

Wood J. 1970. A study of the instantaneous air velocities in a plane behind the wings of certain Diptera flying in a wind tunnel. J. Exp. Biol. 52 (1):17–25.

Wootton R. J. 1976. The fossil record and insect flight. In: Insect Flight. Rainey R. C., ed. Oxford: Blackwell Scientific. pp. 235–54.

————. 1979. Function, homology and terminology in insect wings. Syst. Entomol. 4:81–93.

————. 1981. Support and deformability in insect wings. J. Zool. 193:447–68.

————. 1988. Functional trends in insect wing evolution. In: Proc. 18th Int. Congr. Entomol., Vancouver. 3–8 July 1988. Abstr. and author index. Vancouver. p. 86.

————. 1990. The mechanical design of insect wings. Sci. Am., no. 11:114–20.

————. 1991. The functional morphology of the wings of Odonata. Adv. Odonatol. 5:153–69.

————. 1992. Functional morphology of insect wings. Annu. Rev. Entomol. 37:113–40.

————. 1993. Leading edge section and asymmetric twisting in the wings of flying butterflies (Insecta, Papilionoidea). J. Exp. Biol. 180:105–18.

————. 1995. Geometry and mechanics of insect hindwing fans: a modelling approach. Proc. R. Soc. London B 262:181–87.

————. 1996. Functional wing morphology in hemipteran systematics. In: Schaefer C. W., ed. Studies on Hemipteran Phylogeny. Lanham, Md.: Entomol. Soc. Am. pp. 179–98.

Wootton R. J., Betts C. R. 1986. Homology and function in the wings of Heteroptera. Syst. Entomol. 11 (3):389–400.

Wootton R. J., Ellington C. P. 1991. Biomechanics and the origin of insect flight. In: Rayner J. M. V., Wootton R. J., eds. Biomechanics in Evolution. Cambridge: Cambridge Univ. Press. pp. 99–112.

Wootton R. J., Ennos A. R. 1989. The implications of function on the origin and homologies of the dipterous wing. Syst. Entomol. 14:507–20.

Wortmann M., Zarnack W. 1987. The influence of several parameters of wing movement on the lift of *Schistocerca gregaria* during descending horizontal, or ascending flight. In: Elsner N., Creuzfeldt O., eds. New Frontiers in Brain Research. Stuttgart: Thieme Verl. p. 53.

————. 1993. Wing movements and lift regulation in the flight of desert locusts. J. Exp. Biol. 182:57–69.

Wu J. Z., Vakili A. D., Wu J. M. 1992. Review of physics of enhancing vortex lift by unsteady excitation. Prog. Aerospace Sci. 28:73–131.

Yager D. D., May M. L. 1990. Ultrasound-induced, flight-gated evasive maneuvers in the praying mantis *Parasphendale agrionina*. 2. Tethered flight. J. Exp. Biol. 152:41–58.

Yoshida A., Shinkawa T., Aoki K. 1983. Periodical arrangement of scales on lepidopteran (butterfly and moth) wings. Proc. Jap. Acad. B 59 (7):236–39.

Young J. Z. 1981. The Life of Vertebrates. 3rd ed. Oxford: Clarendon. 645 p.

Zakharenkov M. N., Osminin P. K., Shvets A. I. 1975. The up-to-date state of research of the mechanics of insect flight. In: Trudy Rizhskogo NII Travmatologii i Ortopedii. No. 13, pp. 657–62 (in Russian).

Zalessky G. M. 1943. A brief account of grounds for changing the terminology on insect wings venation. Zool. Zhurn. 22 (3):154–69 (in Russian).

————. 1949. The origin of wings and flight in insects as related to environmental conditions. Uspekhi Sovr. Biologii 28 (3):400–414 (in Russian).

————. 1951. New observations upon the flight of dragonflies. Priroda, no. 3:61–63 (in Russian).

————. 1953. The role of wind in the origin of insect flight. Priroda, no. 11:85–90 (in Russian).

————. 1955. The up-to-date state of the studies of insect flight. Uspekhi Sovr. Biologii 39 (3):308–27 (in Russian).

Zanker J. M. 1988a. How does lateral abdomen deflection contribute to flight control of *Drosophila melanogaster*? J. Comp. Physiol. 162 (5):581–88.

————. 1988b. On the mechanism of speed and altitude control in *Drosophila melanogaster*. Physiol. Entomol. 13 (3):351–61.

——. 1990a. The wing beat of *Drosophila melanogaster*. 1. Kinematics. Phil. Trans. Roy. Soc. London B 327:1–18.

——. 1990b. The wing beat of *Drosophila melanogaster*. 2. Dynamics. Phil. Trans. Roy. Soc. London B 327:19–44.

Zarnack W. 1972. Flugbiophysik der Wanderheuschrecke (*Locusta migratoria* L.). 1. Die Bewegungen der Vorderflügel. J. Comp. Physiol. 78 (4):356–95.

——. 1975. Aerodynamic forces and their calculation in insect flight. In: Wu T. Y., Brokaw C. J., Brennen C., eds. Swimming and Flight in Nature. Vol. 2. New York: Plenum. pp. 797–801.

——. 1983. Untersuchungen zum Flug von Wanderheuschrecken. Die Bewegungen raumlichen Lagebeziehungen sowie Formen und Profile von Vorder- und Hinter-flügeln. In: Nachtigall W., ed. BIONA Report. Stuttgart: G. Fischer Verlag. pp. 78–102.

Zarnack W., Wortmann M. 1989. On the so-called constant-lift reaction of migratory locust. J. Exp. Biol. 147:111–24.

Zavarzin A. A. 1934. On the evolutionary dynamics of tissues. Arch. Biol. Sci. Ser. A. 36 (1):3–64 (in Russian).

——. 1950. Essays on the evolutionary histology of nervous system. In: Selected papers in 4 vols. Vol. 3. Moscow and Leningrad: Izd-vo AN SSSR. 420 p. (in Russian).

——. 1953. Essays of the evolutionary histology of blood and connective tissue. In: Selected papers in 4 vols. Vol. 4. Moscow and Leningrad: Izd-vo AN SSSR. 719 p. (in Russian).

Zavarzin G. A. 1969. Incompatibility of features and the theory of biological system. Zhurn. Obstchej Biologii 30 (1):33–41 (in Russian).

——. 1974. Fenotipicheskaja Sistematika Bakterij. Prostranstvo Logicheskikh Voz-mozhnostej [The Phenotypic Taxonomy of Bacteria. The Space of Logical Oppor-tunities]. Moscow: Nauka. 141 p. (in Russian).

Zherikhin V. V., Gratshev V. G. 1995. A comparative study of the hind wing venation of the superfamily Curculionoidea, with phylogenetic implications. In: Pakaluk J., Slipinski S. A., eds. Biology, Phylogeny, and Classification of Coleoptera: The papers celebrating the 80th birthday of Roy A. Crowson. Vol. 2. Warsaw: Muzeum i Instytut Zoologii PAN. pp. 634–777.

Zimov S. A. 1993. Azbuka Risunkov Prirody [The Alphabet of Patterns in Nature]. Moscow: Nauka. 125 p. (in Russian).

Index

Page numbers in bold refer to illustrations; in italics, to tables

Acantholyda nemoralis, 123, 152
Acmaeodera, 103
Acmaeoderella, 103
adaptedness, 176
adaptive syndrome, 192
adaptive trade-offs, 197–9
additivity, 180–5
Adelidae, 160
adequacy, 170–1, 200–4
advance ratio, 73, **74**
Aedes, 145, 147; *cyprius,* **133**
Aegeria, 104
Aegeriidae, **89,** 109, 167
aerodynamics: aircraft, 22–5, 54–5;
  flapping flight, 23–7, 35–61, 73, 113–6
Aeshna, 55
Agathiphagidae, 135, 143
Agathomyia viduella, 148
Aglais urticae, **63,** 69
Agriphila straminella, **131**
Agrypnia pagetana, **130**
Alberch, P., 172
alderflies. *See* Corydalida
Alexander, R. McN., 73, 187, 200
Allancastria cerisyi, 158
Allantus, **89,** 99
allogenesis, 192, **193**
allometry, 187–8, 201–2
Allonarcys sacchalina, 15
Alloperla mediata, **88**
Alphus sylvinus, 132
Alsophila pometaria, 108
Alucitidae, 108
alula, 112, 118, 147

Amblyrrhina maculata, 99
Ammophila, **89**
Amphiesmenoptera, 104–9, 125–43
amphisbaenians, skull morphology of,
  189, **190**
amplitude, stroke, 33, 67–71, 104
Amuriella: dieckmanni, 161–2; *fulgens,* 162
Anabolia, **89;** *laevis,* **122, 140,** 165
anal loop, 135, 173
analogy, 119, 179–81
Ancistrocerus, **89;** *parietinus,* **125**
andean condor, 67
angle of attack, 18, 20, 55–6, 70, 72–3,
  112–5, 187
Anisopus, **124,** 147
ant lions. *See* Myrmeleontida
anteromotorism, 12–5, 35–42, 48–51, 86,
  96, 117
Antichaeta, 148
ants. *See* Vespida
Apatania, **89,** 105, **116;** *wallengreni,* **122,**
  141
Apis mellifera, 69, **89, 125**
archetype, 177–8
Archips podana, **138**
Archon, 157
Arctiidae, 139, 143
Arctotypus, **6**
arculus, 138
Ardea cinerea, **186**
Ardeola ralloides, **186**
Arge, **89**
Arguda vinata, 161–2
Arichanna melanaria, **164**

arogenesis, 192, **193**
aromorphosis, 192
Asilidae, **89**, 147
Asilus, **124**
aspect ratio, **91**, 97
asymmetry of strokes, 17–8, 75, 87
asynchronous muscles, 109
axillary cord, 137
axillary sclerites, 121, 126–7, 142, 147, 151–2

Baicalini, 12
bat flight, 68, 73
bees. *See* Vespida
beetles. *See* Scarabaeida
Belostomatidae, 98
Beloussov, L. V., 178
Bharetta cinnamomea, 162
Bhutanitis thaidina, 158
Bibio, **89**
bird flight, 67, 73–4, 115, 186–7
Blattella germanica, 84
Blattida, 10, 15
Blattodea. *See* Blattida
Blattogryllus caratavicus, **11**
Bocharova-Messner, O. M., 41, 57, 64, 122, 168
body angle, 68
Bohemannia ussuriella, **133**
Bohr, N., 183
Bojophlebia prokopi, **5**
Boloria dia, **164**
Bombus, **89**, *terrestris*, 68
Bombycoidea, 160
Bombyliidae, 147
booklice. *See* Psocida
bound vortex. *See* circulation
boundary layer, 57, 146, 168
Bovidae, 202–3
Brachycentrus subnubilus, 136, 141
Brachycera, 112, 116, 147
Brodsky, A. K., 12, 15, 41, 64, 78, 168
bugs. *See* Cimicida
bullae. *See* fenestrae
bumblebees. *See* Bombus
Buprestidae, 103
butterflies. *See* Papilionida

caddisflies. *See* Phryganeida
Calamoceratidae, 139, **140**
Calliphora: *erythrocephala*, 68, **77**; *vicina*, 69
Calliphoridae, **89**

Calopteryx splendens, 55
calypteres, 147
Cantharididae, **89**
Cantharis, 43–7
Capperia, 139
Cassida, 151
Catocala nupta, 164
Celastrina argiolus, 166
Ceraclea nigronervosa, 141, 165
Cerambycidae, **89, 102**
Ceratopogonidae, 146
Cerion, 172
Ceropales, 152
Cetonia, **89**, 101, 103; *aurata*, 151
Cetoniinae, 13
Chaetopteryx villosa, 121, **140**, 165
Chai, P., xiii, 71
character, 177
Chilena sordida, 162–4
Chironomidae, **133**, 146
Chlorophorus faldermanni, 150
Choreutidae, 143
Chrysis, 152
Chrysopa, **89**, 115; *carnea*, 88; *dasyptera*, 48–51, 88, **128**
Chrysotoxum, 145
Churchill, W., 198
Cicindela, 14, 43–7, 86, 101, **102**, 123; *restricta*, 151
Cimicida: in-flight vortex generation, 37, 40; taxonomic relationships, 8; wing deformation, 57, 116; wing morphology, 135–6, 146–7; wingbeat kinematics, 12–4, **34**, 68, 98–101, 114
Cimiciformes, 4, 97
circulation, 23, 45, 79, 93, 115
clap, 87, 103
clap-and-peel, near, 99, 109, 112
clavus, 20, 141, 150–1
Coccinella trifasciata, 151
Coccinellidae, **89**
Coccoidea, 13, 99
cockroaches. *See* Blattida
Coleoptera. *See* Scarabaeida
color patterns, **157**, 165–6, 171–3, 195
combined vortex. *See* vortices: U-shaped
complementarity in biology, 180–5
Comstock, J. H., 126, 137
coot, 68
concentration of function, 116
convergence, 119, 179–81
corium, 116
Corydalida, 9, 15, 87, **89, 91**

Corymbia rubra, **102, 123,** 151
Cosmotriche lunigera, 162
Cossidae, **89,** 158
costalization, 153
Crambidae, 143
crane flies. *See* Tipulidae
crickets. *See* Gryllida
Culicidae, **89,** 169
Culicoides grisescens, **133**
Curculionidae, **89,** 169

damselflies. *See* Libellulida
D'Arcy Thompson, W., 191, 196, 201
Dasyleptus brongniarti, **4**
Dasyneura laricis, **133**
deformation. *See* wings: in-flight deformation
Dermaptera. *See* Forficulida
Deuterophlebiidae, 148, **149,** 151
developmental constraints on evolution, 170, 201
Diaphanopterida, 7
Dickinson, M. H., xiv, 42, 56, 66, 71, 79
Dictyoneurida, 6
Dipluriella loti, 162
Diptera. *See* Muscida
dipterous flies. *See* Muscida
direct flight muscles, 97
Discodes aeneus, **133**
Dogel', V. A., 116
Dolichovespula silvestris, **125**
downstroke-to-upstroke duration ratio, **18,** 31, 68
dragonflies. *See* Libellulida
Drosophila, 42, 56, 66, 68–9, 75, 79, **89, 133;** *hydei,* 69; *melanogaster,* 69, 71; *virilis,* 55, 69
Dudley, R., xii–xiv, 82, 95, 109

earwigs. *See* Forficulida
ectothermy, 104–8
Ellington, C. P., xiv, 12, 56–8, 64
elytra: defensive function of, 99, 101; flow around, 45; in-flight motion of, 14–5, 43–5, 101
Embiida, 10, 15, 87, 94
Embioptera. *See* Embiida
Emelianov, A. F., xiii, 126, 129
Encarsia formosa, 51, 64, 99
Endopterygota. *See* Scarabaeiformes
endothermy, 104–9, 167
energy accumulation, 79

Ennos, A. R., 145, 148
Entomoscelis adonidis, 150
Ephemerida, 4, 13, 81, 117
Ephemeroptera. *See* Ephemerida
Epistrophe, 145
Eriocrania semipurpurella, 160
Eriocraniidae, 15, 87, **89,** 104, 135, 157, 160, 165
Eriogaster: catax, 162–3; *lanestris,* 160, 162
Eriozona, 145
Eristalis, 145
Eulonchoptera, 152
Eustaudingeria vandalica, 162
evolution, 176, **205**

faculty, definition of, 177
fairy moths. *See* Adelidae
falcon, 69–70
fanlike air throw-off, *39,* 41, 60–1, 64, 72, 79
feather-like wings, 108
feature, 177
feeding-to-flight correlation, 117–8
fenestrae, 138, 147, 152, 154
flight: adaptive radiation, 90–4; flapping, xii, 23–7, 58–9; flitting, 12; free, 31, 34; gliding, 117; mathematical models of, 53–4, 58, 60–1, 77–8; mechanical models of, 54, 58; origin of, 82–6; straight, 12, 19, 47, 73; tethered, 29–35, 42, 67–70. *See also* anteromotorism; hovering; posteromotorism
fling, 51–4, 59–61, 63–4, 79, 87, 113–4
flow visualization, 27–9
Forficulida, 10, 14, 148, **149**
four-wingedness: functional, 14–5, 87, 101; morphological, 13
frequency, wingbeat, 14, 67–71, **89,** 97, 106, 109, 148, 155
function, 176–7
furrows: absence of, 115; ano-jugal, 135–7, 141–2, 144, 147, 150–1; claval, 141–2; functioning at folding, 103, 110, 139, 141–3, 148, 152; functioning during flight, 99, 103, 151; general, 139; morphology, 123, 151, 153–4; nomenclature, 20, 129; peripheral, 139, 145, 154; primary, 20, 138, 147; remigial, 139, 143–4, 147, 151–2; remigio-anal, 103, 135, 139, 141–4, 147, 150, 152; secondary, 141–3, 152; topology of, 120, **138, 140;** types of, 153–4

Ganonema extensum, 136, **140**
Gans, C., xiii–xiv, 170, 172–3, 200–4
Gastropacha: populifolia, 162; *quercifolia,*
    161, 163–4
gear change mechanism, 112–3
Geometridae, 108
Geometroidea, **89**, 160, 164
German cockroach. *See Blattella ger-
    manica*
ghost moths. *See* Hepialidae
Glyphotaelius pellucidus, **140**, 165
Goeridae, **140**
Goetz, K. G., 42, 56, 66
Gonepteryx rhamni, 41, **134**
Graphium sarpedon, 68
grasshoppers. *See* Gryllida
Gryllida, 10, 15, 20, 32, 67–70, 81, 93–4
Grylloblattida, 10, 94
Gryllones, 10, 94–6, 118
Gryllotalpidae, 14

halteres, 93, 110
Hamilton, K. G. A., 126–7
Harris' hawk, 69
haustellate mouthparts, 117
Heinrich, B., 106–9, 166–8
Heliconiidae, **89**
Heliconius, 56
Heliocopris, **89**
Heliothis, 116; *armigera,* **33, 105**
Helophilus, 145
Hemerobius, **89**
Hemiptera. *See* Cimicida
Hepialidae, 104, 126, 132, 160
herons, 186–7
Hesperioidea, 106, 160, 164
heterarchy, 208
Heterobathmiidae, 135
hierarchy, 185
Hippoboscidae, 147
Holometabola. *See* Scarabaeiformes
homogeny, 173, 178–9
homoiology, 179–80
homology, 172–3, 178–9
homoplasy, 119, 143, 171–2, 179
honeybee. *See Apis mellifera*
hovering, 12, 45, 74–5
hummingbirds, 69, 71
Hutchinson, G. E., 196
Hybomitra pavlovskii, **124,** 145
Hydropsyche nevae, **122, 130**
Hydroptila, **89;** *vectis,* **133**
Hymenoptera. *See* Vespida

ibises, 187
Ichneumonidae, **89**
idioadaptation, **193**
Illiberis sinensis, 121, **131, 138**
Inachis io, **63,** 77
inadaptation, 198
industrial melanism, 171
innovations, origin of, 116, 172–3
Iphiclides, 158; *podalirius,* **156**
Irish elk. *See Megaloceros giganteus*
Ischirosyrphus, 145; *glaucius,* **124,** 145
Isogenus nubecula, 15
isomorphism, 179–80
Isoptena serricornis, 15
Isoptera. *See* Termitida

jugum, 136–7, 147, 151–2

Kalligrammatidae, 172
Kantenschwung, 20, 41, 61, 114–5
kinematics, wingbeat: antiphase, 16,
    96–7, 110; classification of, 11–6; func-
    tional aspects, 113–6; during horizon-
    tal turns, 95, 97, 113; initial condition,
    86–8; in-phase, 15; quasisynchronous,
    15
Kofman, G. B., xii, 181, 183, 201
Kokshaysky, N. V., xiii, 74, 116, 186–92
Kukalova-Peck, J., 150

lacewings. *See* Myrmeleontida
Lagria, 151
Laothoe populi, **164**
lappet moths. *See* Lasiocampidae
large white. *See Pieris brassicae*
Lasiocampidae, 158, 160–4
Lasiocampoidea, 163
Laspeyresia pomonella, **30, 32, 131**
Lehmann, F.-O., 42, 71
Lepidoptera. *See* Papilionida
Lepidostoma hirtum, **130**
Leptidea sinapis, **62,** 116
Leptogaster cylindrica, 110, **111**
Leptogastridae, 147
Lethe eumolphus, 166
Leucorrhinia rubicunda, 98
levels, structural and functional, 185,
    200
Liassophilidae, 113
Libellulida, 4, 15–6, 35, 55, 62, 67–70, 81,
    83, 96–8
lift, 23; regulation, 69–70; within-stroke
    distribution, 62–6

lift-control reaction, 32, 95
lift-to-drag ratio, 55
Limacodidae, 143
Limnephilidae, **122, 140,** 165
line, flexion and fold. *See* furrows
Lipoptena cervi, 123, **124,** 147
locust, desert. *See Schistocerca gregaria*
locust, migratory. *See Locusta migratoria*
Locusta migratoria, 68–70
Lucanus, **89**
Luehdorfia puzioli, 158
Lymantriidae, 143

magpie, 68
Malacosoma: castrensis, 162; *franconica,*
     162–3; *neustria,* 162–3
Mamkaev, Yu. V., 192
mandibulate moths. *See* Micropterigidae
mandibulate mouthparts, 117
Manduca sexta, 58, 77
Manteida, 10, 20
mantids, praying. *See* Manteida
Mantodea. *See* Manteida
Margaritia sticticalis, **138**
Martynov, A. V., 2, 129, 137
Maslov, S. P., xiii, 197
mayflies. *See* Ephemerida
Mecoptera. *See* Panorpida
median plates, 127
Megaloceros giganteus, 171
Megaloprepus coerulatus, 69–70
Megaloptera. *See* Corydalida
Megasecoptera. *See* Mischopterida
Melanthus, 166
Melolontha, **89**
membrane (wing zone in Cimicida), 116
merone, 178
Metanastria hyrtaca, 162–4
Metrocampa margaritata, **134**
Meyen, S. V., 177–8, **193,** 196
Micropterigidae, 15, 87, **89,** 104, 157, 160,
     165
Micropterix calthella, 160
miniaturization. *See* scale effects
minimum change in evolution, 191–2
Miomoptera. *See* Palaeomanteida
Mischopterida, 7
Molanna: albicans, 135; *angustata,* 135
mole crickets. *See* Gryllotalpidae
Monochamus: sutor, 100, 151; *urussovi,*
     102, 151
Mordellistena, **89;** *tournieri,* **102,** 151

morphogenetic factors in evolution,
     172–3, 181–4, 206
morphology: comparative, 178; func-
     tional, xi, 184–5
morphospace, 179; filling of, 192–6
moths. *See* Papilionida
mountain midges. *See* Deuterophlebii-
     dae
multifunctionality, 155, 197, 203
multivariate correspondence, structures-
     to-functions, 154, 185, 197–200
Musca domestica, 67, 69
Muscida: in-flight force generation, 62,
     66–71; in-flight vortex generation, 37–
     8, 40, 55–6, 60, 93; morpho-functional
     parameters, **89–91,** 110–5; taxonomic
     relationships, 9; wing morphology,
     144–8; wingbeat kinematics, 13, 95
Muscidae, **89**
Mymaridae, 13, 99
Mymarommatidae, 13, 99
Myrmeleon, **89**
Myrmeleontida, 9, 13, 15, 43, 48–51, 87,
     89–91, 143–4, 172

Nachtigall, W., 20, 47, 66, 114, 168
Nadiasa undata, 162
Narosoideus flavidorsalis, 121
natural selection, xi, 116, 146, 153, 175,
     180
Necrophorus, 103
Nemapogon dorsiguttella, **131**
Nemopteridae, 13, 113
Nephrotoma, **124**
Neuroptera. *See* Myrmeleontida
nightingale, 67
Noctuidae, **89,** 143, 165
Noctuoidea, 106, 163, 167
nodal line, 116, 138, 151–2, 154
Notodontoidea, 106, 163
novelties, 116, 172–3
Novokshonov, V. G., 121–2
Nymphalidae, 156, 160, 167

Odagmia ornata, **133**
Odonata. *See* Libellulida
Odonestis pruni, 162
Oedemera, 151
Oidaematophorus, 139
oligomerization, 153
Oligoneoptera. *See* Scarabaeiformes
Oligotricha striata, 165

Operophtera bruceata, 108
optimality, 144, 200–4
Ornithoptera, 14, 158
Orthoptera. *See* Gryllida
Oxyporus, 122, **123**

Pachygaster atra, 123, **124**
Palaeodictyoptera. *See* Dictyoneurida
Palaeomanteida, 8
Palaeomantina pentamera, **8**
Panorpa, 115; *communis,* **88, 128,** 144;
    germanica, 88
Panorpida, 9, 15, 87, **89–91,** 113, 126,
    143–4
Paoliida, 3, 6, 121, 126, 144
Papilio, 14; *maacki,* **156;** *machaon,* **156;** *or-*
    menus, 166; *penelope,* 166; *rumanzovia,*
    68; *ulysses,* 166; *ybecatheus,* 166
Papilionida: aerodynamic force gen-
    eration, 61–5, 67–70, 113–4; in-flight
    vortex formation, 35, 37, 40–1, 55, 75,
    77; morpho-functional parameters,
    89–91, 104–9; taxonomic relationships,
    9; wing deformation, 20; wing mor-
    phology, 126–43, 155–69; wingbeat
    kinematics, 13–5, **30, 32–3,** 86
Papilioninae, 158
Papilionoidea, **89,** 106, 160, 164
Parachiona picicornis, 165
Paralebeda plagifera, 161
parallelism, 179
Paraneoptera. *See* Cimiciformes
paranota, 83–6, 117
Parides, 158; *neophilus,* 14
Parnassiinae, 158
Parnassius, 156, **157,** 165, 167
peacock butterfly. *See* Inachis io
peel, 37, 57, 103, 109, 114
Perlida, 10, 12, 15, 82, 87, 94
Permopsychops saurensis, 122
Permotanyderidae, 113
Pfau, H. K., 112
Phagmatobia fuliginosa, **138**
phase shift, 70
Phasmatida, 10, 136
Phasmatodea. *See* Phasmatida
Philudoria: albomaculata, 162; *potatoria,*
    162
Phormia regina, 66
Phryganea bipunctata, **130, 140,** 165
Phryganeida: in-flight vortex gen-
    eration, 37, 51; morpho-functional
    parameters, **89–91,** 104–10; taxonomic

relationships, 9; wing morphology,
    121, 126–42, 163, 169; wingbeat kine-
    matics, 12, 14–5
Phyllodesma: alice, 162; *glasunowi,* 162;
    ilicifolia, 162–3; *suberifolia,* 162
Pielus labirinthecus, **127**
Pieridae, 106, 160, 167
Pieris brassicae, **65,** 75
pigeon, 68
Platypezidae, 147
Plecia, 147
Plecoptera. *See* Perlida
plica. *See* furrows
Poecilocampa populi, 162–3
Polygonia c-album, **63**
polymerization, 153, 155
Polyneoptera. *See* Gryllones
Polyommatus: amandus, 166; *coridon,* 166;
    daphnis, 166
Pompilidae, 152
Ponomarenko, A. G., 148
positional angle of wing, **18-9,** 49, 59
posteromotorism, 12–5, 43–8, 93, 96, 117
Posthosyrphus, 145
power regulation, 67–71
preadaptation, 172
precubital thickening, 121–2, 145
preflight warm-up, 108
Pringle, J., 47
progress, evolutionary, 192–4
pronation, 19, 61, 70, 77, 79, 87, 103, 116,
    148
proto-adaptation, 172–3
Pseudogonalos hahni, 152
Psocida, 7
Psocoptera. *See* Psocida
Psychidae, 165
Psychoda, **133**
Psychopsidae, 172
Pterophylla camellifolia, 86
Pterygota. *See* Scarabaeona
Ptychoptera lacustris, **124,** 145
Ptychopteridae, 145
punctuated equilibria, 198
pygmy grasshoppers. *See* Tetrigidae
pygmy mole crickets. *See* Tridactylidae
Pyraloidea, **89, 138,** 143, 160, 164

Raphidiida, 9, 15, 87, **91**
Raphidioptera. *See* Raphidiida
Rashevsky, N., 191
Rasnitsyn, A. P., xii–xiv, 1, 3, 84–6, 153,
    197–8, 207

regress, evolutionary, 193, 194
remigium, 20, 103, 141, 150–1
repeated polymorphic sets, 196
Reynolds number, 55–6, 58, 116, 168
Rhogogaster, **125,** 153
Rhyacophila, **89;** *nubila,* 127, 140; *tristis,* **140;** *vulgaris,* **140,** 141
Rhyacophilidae, 15, 87, 104
roaches. *See* Blattida
Rohdendorf, B. B., 3, 12, 108
Ruppell, G., 96–8

salamander foot, evolution of, 172
Saturniidae, **89,** 106, 172
Saturnioidea, **89**
Satyridae, 156, 160, 167
scale covering, 57, 115, 137, 155–69; functions of, 166–9; heat-insulating capacity of, 108–9, 166–8; layers of, 158–66; main types of, **161**
scale effects, 66, 108, 134–5, 139, 145–6, 168, 186–9
scale insects. *See* Coccoidea
scales: aerodynamic significance, 168; morphology, 155–6, 164–6; orientation on wing, 156, 165; origin, 169, 173; rows of, 157–8
Scarabaeida: aerodynamic force generation, **34,** 68; in-flight vortex formation, 43–8, 57; morpho-functional parameters, **89–91,** 101–4; taxonomic relationships, 8; wing morphology, 146, 148–51; wingbeat kinematics, 14–5, 118
Scarabaeidae, **89**
Scarabaeiformes, 8, 97, 150
Scarabaeona, 3, 82; ancestors of, 83, **85,** 116–7
Schistocerca gregaria, 55, 69
Schmalhausen, I. I., 94, 191, 199–201
Schmidt-Nielsen, xii, 189, 202
scorpion flies. *See* Panorpida
secondary signals, 189
Semblis, **89**
separation of airflow, 22–4, 36, 44, 48, 51–8, 113–4, 186–9
Serecinus telamon, 158
Severtzov, A. N., 177, 192–5, 197
Sharov, A. G., 3
Shwartz, S. S., 189–91
Sialida. *See* Corydalida
Silo nigricornis, **140,** 141
similarity, 186, 202–3

Simuliidae, **89,** 146
small tortoiseshell. *See Aglais urticae*
snakeflies. *See* Raphidiida
soldier beetles. *See Cantharis*
speed of flight, 33, 67–9
Sphaerophoria taeniata, **111**
Sphecidae, **89**
Sphegionoides, 145
Sphingidae, **89,** 106, 156, 165, 167. *See also Manduca sexta*
Sphingoidea, 106, 163, 167
stabilization of functions, 186–92
stick insects. *See* Phasmatida
Stigmella kozlovi, **133**
stoneflies. *See* Perlida
storks, 187
Stratiomyidae, 147
Strepsiptera. *See* Stylopida
stroke angle. *See* amplitude, stroke
stroke phases, 61
stroke plane: horizontal, 12; inclined, 12, 19, 97, 110; position of, 12, 17, 68–70; vertical, 12, 104
structuralism, 182–3
Stylopida, 8, 14
supination: associated vortices, 37, 39, 73; part of stroke cycle, 19–20, 61, 64, 148; in useful force generation, 70, 75, 77, 79, 115–6
surface structures, 57, 146, 154. *See also* scale covering; scales
Svidersky, V. L., xiii
swinging-edge mechanism. *See* Kantenschwung
symmorphosis, 98, 200
Sympetrum sanguineum, 55
Synomaloptila longipennis, **7**
Syntomidae, 13
Syntomis phygea, **138,** 142
Syrastrenopsis moltrechti, 162
Syrphidae, **89,** 121, 145
Syrphus, 145; *balteatus,* 55

Tabanidae, **89**
Tachina magnicornis, **124,** 145
Taeniopteryx nebulosa, 15
taxon: definition of, 176; maturation of, 195
taxonomy, insect: current, 3–11; history, 1–3
Temnostoma, 145
temperature of pterothorax, 33, 166
Tenthredo, **89**

Tephrochlaena, **111**
termites. *See* Termitida
Termitida, 10
Tetrigidae, 14
Thaumetopoea: herculeana, 161; *iordana,* 162–3; *pityocampa,* 162; *solitaria,* 162–3
Thaumetopoeidae, 160
thermoanemometry, 42, 71
Thripida, 7, 146
thrips. *See* Thripida
thrust, 62–4, 112
Thymelicus lineola, **30**
thyridium, 138, 144, 148
Thysanoptera. *See* Thripida
tiger beetle. *See* Cicindela
tip vortex, 24, 49, 56, 76
Tiphia femorata, **125,** 152
Tipula oleracea, 55, 110, **111**
Tipulidae, 41–3, 55, **89,** 112
Tipulomorpha, 110
tobacco hawkmoth. *See Manduca sexta*
Tortricidae, 139, 143
Trabala vishnou, 162
trait, 177
trajectory: of flight, 12; of wings, 17–9, 70, 73, **77**
trapezia-like deformation, 104, 109, 115, 139
Triaenodes bicolor, 135
Triatoma infestans, 69–70
Trichiura crataegi, 162–3
Trichius, **89,** 122; *fasciatus,* **123,** 151
Trichoptera. *See* Phryganeida
Tridactylidae, 14
Trigona, **89**
Troides radamantus, 69
Tschekardobia osmylina, **7**
twisted-winged parasites. *See* Stylopida
two-wingedness, 93; functional, 14; morphological, 13, 99; vorticity at, 35–42

Uldrick, J. P., 26, 59
uncertainty of explanations in morphology, 184
ungulates, 187–9
Urocerus gigas, 99, *100,* 116, 123, **125,** 152

vannus, 141–2
veins: anal, 127–9, 135–7, 142; corrugated, 121–3, 145–6, 173; costa, 99, 146, 148; cubitus anterior, 121, 126,

145, 150, 206; cubitus posterior, 121, 126, 129, 135, 138, 145; empusal, 127; false, 121, 145, 153, 173; jugal, 136–7, 153; media, 20, 129, 138, 145, 148, 150; nomenclature of, 126; postcubitus, 127, 142; radial sector, 121; radius, 121, 206; reduction of, 132–5; secondary, 121; subcosta, 121; tracheation of, 126
vena arcuata, 136–7
vena cardinalis, 136–7
vena spuria, 121, 145
venation, proximo-distal shift of, 153
Vermej, G. J., 95–6, 198, 202
vertebrate limbs, 170, 172–3, 182, 187–9
vertical oscillations: of abdomen, 64–6; of thorax, 61–4
Vespa, 89
Vespida: aerodynamic force generation, **34,** 67–70, 114; in-flight vortex formation, 51, 55; morpho-functional parameters, **89–91,** 98–101; taxonomic relationships, 9–10; wing morphology, 136, 146, 151–3, 169; wingbeat kinematics, 13–4, 35
vortex rings, 23, **24,** 37–42, 51, 72, **74;** chain-linked 74, **76;** coalescence of, 77
vortex tube, 23, 59, **74,** 78
vortex wake, 25–8, 35, 71–9; longevity of, 78
vortices: conical spiral shaped, 56; dorsal, 37, 43–4, 49, 76; in free and tethered flight, 34; interaction of, 86–7; starting, 23–6, 36–7, 43, 49, 51, 72–3, 90, 94, 114, 117; stopping, 23, 39, 42–3, 51, 59, 72–3; study of, 27–9; U-shaped, 26, 37, 43, 49–51, 59–60, 72, 75, 79, 115; ventral, 40, 43. *See also* tip vortex

Wagner's effect, 58, 66
Wake, D. B., 192
wasps. *See* Vespida
web spinners. *See* Embiida
Weis-Fogh, T., 51, 53, 59
Willmott, A. P., 41, 56–8, 64
wing: ano-jugal zone, 103; functional, 12; lanceolate shape, 132; loading, 91, 101, 103, 108; morphological, 12; pair, 12; planform, 106, 110, 129–30, 143–5
wings: couple, 12, 93, 117; coupling mechanisms, 104, 118, 137; in-flight deformation, 20, **21,** 70, 99, 103, 109–

ᵉᵉcᵢ

cᵢᵢ

10, 144; fore and hind, relative size of, 12–3, 93, 99, 118; functions, 81, 116–7, 144; precursors, 82–4; ventral flexion, 99, 104, 109, 112, 115–6, 139, 153
Wootton, R. J., xiv, 64, 129, 145, 148

Xylocopa, **89**
Xylotrechus namanganensis, 150

Zdenekia grandis, **5**
Zerynthia, 156
Zherikhin, V. V., xiii, 113, 204
Zophomyia temula, 148
Zoraptera. *See* Zorotypida
Zorotypida, 6, 87
Zygaenidae, **89**

Library of Congress Cataloging-in-Publication Data

Grodnitsky, Dmitry L., 1962–
    Form and function of insect wings : the evolution of biological
structures / Dmitry L. Grodnitsky.
        p.   cm.
    Includes bibliographical references (p.   ) and index.
    ISBN 0-8018-6003-2 (alk. paper)
    1. Insects—Flight.   2. Insects—Morphogenesis.   I. Title.
QL496.7.G76   1999
595.7147'9—dc21   98-30339
                                                                CIP